Statistics in Music Education Research

Statistics in Music Education Research

A REFERENCE FOR RESEARCHERS, TEACHERS, AND STUDENTS

JOSHUA A. RUSSELL

UNIVERSITY PRESS

OXFORD
UNIVERSITY PRESS

Oxford University Press is a department of the University of Oxford. It furthers
the University's objective of excellence in research, scholarship, and education
by publishing worldwide. Oxford is a registered trade mark of Oxford University
Press in the UK and certain other countries.

Published in the United States of America by Oxford University Press
198 Madison Avenue, New York, NY 10016, United States of America.

© Oxford University Press 2018

All rights reserved. No part of this publication may be reproduced, stored in
a retrieval system, or transmitted, in any form or by any means, without the
prior permission in writing of Oxford University Press, or as expressly permitted
by law, by license, or under terms agreed with the appropriate reproduction
rights organization. Inquiries concerning reproduction outside the scope of the
above should be sent to the Rights Department, Oxford University Press, at the
address above.

You must not circulate this work in any other form
and you must impose this same condition on any acquirer.

Library of Congress Cataloging-in-Publication Data
Names: Russell, Joshua A. author.
Title: Statistics in music education research : a reference for
researchers, teachers, and students / Joshua A. Russell.
Description: New York, NY : Oxford University Press, [2018] |
Includes bibliographical references and index.
Identifiers: LCCN 2017026158 (ebook) | LCCN 2017027559 (print) |
ISBN 9780190695231 (updf) | ISBN 9780190695248 (epub) |
ISBN 9780190695217 (alk. paper) | ISBN 9780190695224 (alk. paper)
Subjects: LCSH: Music—Instruction and study—Statistics.
Classification: LCC MT1 (print) | LCC MT1 .R89 2018 (ebook) | DDC 780.72/7—dc23
LC record available at https://lccn.loc.gov/2017027559

9 8 7 6 5 4 3 2 1

Paperback printed by WebCom, Inc., Canada
Hardback printed by Bridgeport National Bindery, Inc., United States of America

For Melissa, Anna Claire Elise, and Emmeline Quinn

CONTENTS

Preface xv
About the Companion Website xxi

SECTION I | Introduction 1

1. Fundamental Principles 3
    Hypothesis Testing 3
    Measures of Central Tendency 8
    Measures of Variability 14
    Error 18
    Types of Variables 20
    Types of Data 20
    Types of Statistical Procedures 22
    Significance 24
    Degrees of Freedom 25
    Moving Forward 29

2. Descriptive Statistics 31
    Use of Descriptive Statistics 31
    Assumptions of Descriptive Statistics 34
    Research Design and Descriptive Statistics 34
    Setting Up a Database for Descriptive Statistics 35
    Examining the Normality of Descriptive Statistics 36
    Analyzing Nominal or Categorical Descriptive Statistics 38
    Analyzing Ratio or Scaled Descriptive Data 39
    Reporting Descriptive Statistics 42

Writing Hints for Descriptive Statistics 42
Final Thought on Descriptive Statistics 43
Descriptive Statistics Exercises 43

SECTION II | Parametric Statistical Procedures 45

3. Pearson Product-Moment Correlation 47
   Use of Pearson Product-Moment
      Correlation Tests 47
   Assumptions of Pearson Correlation 50
   The Null Hypotheses for the Pearson Product-
      Moment Correlation 51
   Research Design and Pearson Correlation 52
   Setting Up a Database for a Pearson Correlation 52
   Conducting a Pearson Correlation in SPSS 52
   Final Thought on Pearson Correlation 56
   Reporting a Pearson Correlation 57
   Writing Hints for a Pearson Product-Moment
      Correlation 57
   Pearson Product-Moment Correlation Exercises 58

4. One-Sample T-Test 59
   Use of the One-Sample T-Test 59
   Assumptions of One-Sample T-Tests 60
   Research Design and the One-Sample T-Test 61
   The Null Hypothesis for the One-Sample T-Test 62
   Setting Up a Database for the One-Sample
      T-Test 63
   Conducting a One-Sample T-Test in SPSS 63
   Reporting a One-Sample T-Test 65
   Writing Hints for a One-Sample T-Test 65
   One-Sample T-Test Exercises 66

5. Dependent-Samples T-Test 67
   Use of the Dependent-Samples T-Test 67
   Assumptions of the Dependent-Samples T-Test 69
   Research Design and Dependent-Samples T-Test 70
   The Null Hypothesis for a Dependent-Samples
      T-Test 70
   Setting Up a Database for a Dependent-Samples
      T-Test 71
   Conducting a Dependent-Samples T-Test in SPSS 71

Reporting a Dependent-Samples T-Test   77
Writing Hints for Dependent-Samples T-Test   77
Final Thought of Dependent-Samples T-Test   78
Dependent-Samples T-Test Exercises   78

6. Independent-Samples T-Test   79
   Use of the Independent-Samples T-Test   79
   Assumptions of the Independent-Samples T-Test   80
   Research Design and Independent-Samples T-Test   82
   The Null Hypothesis for an Independent-Samples T-Test   84
   Setting Up a Database for an Independent-Samples T-Test   85
   Conducting the Independent-Samples T-Test in SPSS   85
   Reporting an Independent-Samples T-Test   90
   Writing Hints for Independent-Samples T-Tests   90
   Final Thought on Independent-Samples T-Test   91
   Independent-Samples T-Test Exercises   92

7. Univariate Analysis of Variance (ANOVA)   93
   Use of the Analysis of Variance   93
   Post Hoc Tests for ANOVA   94
   Assumptions of Univariate ANOVA   95
   Research Design and the Univariate ANOVA   97
   The Null Hypothesis for the Univariate ANOVA   100
   Setting Up a Database for the Univariate ANOVA   100
   Conducting a Univariate ANOVA in SPSS   101
   Reporting a Univariate ANOVA   109
   Writing Hints for Univariate ANOVA   110
   Univariate ANOVA Exercises   111

8. Factorial Analysis of Variance (ANOVA)   113
   Use of the Factorial ANOVA   113
   Post Hoc Tests for Factorial ANOVA   115
   Reporting Factorial Analysis of Covariance in Table Format   116
   Assumptions of the Factorial ANOVA   117
   Research Design and the Factorial Analysis of Covariance   118

        The Null Hypotheses for the Factorial ANOVA   118
        Setting Up a Database for the Factorial ANOVA   120
        Conducting a Factorial ANOVA in SPSS   120
        Reporting a Factorial ANOVA   126
        Writing Hints for Factorial ANOVA   127
        Factorial ANOVA Exercises   128

9. Multivariate Analysis of Variance (MANOVA)   129
        Use of the Multivariate Analysis of Variance
          (MANOVA)   129
        Assumptions of MANOVA   131
        Research Design and the MANOVA   134
        The Null Hypotheses for MANOVA   135
        Setting Up a Database for a MANOVA   136
        Conducting a MANOVA in SPSS   136
        Reporting a MANOVA   144
        Writing Hints for MANOVA   144
        MANOVA Exercises   146

10. Repeated-Measures Analysis of Variance   147
        Use of Repeated-Measures ANOVA   147
        Assumptions of the Repeated-Measures
          ANOVA   150
        Research Design and the Repeated-Measures
          ANOVA   151
        The Null Hypotheses for the Repeated-Measures
          ANOVA   152
        Setting Up a Database for a Repeated-Measures
          ANOVA   153
        Conducting a Repeated-Measures ANOVA in
          SPSS   153
        Reporting a Repeated-Measures ANOVA   159
        Writing Hints for Repeated-Measures ANOVA   159
        Repeated-Measures ANOVA Exercises   160

11. Univariate Analysis of Covariance (ANCOVA)   163
        Use of the Analysis of Covariance   163
        Post Hoc Tests for ANCOVA   165
        Assumptions of ANCOVA   165
        Research Design and the ANCOVA   166
        The Null Hypothesis for the ANCOVA   168
        Setting Up a Database for the ANCOVA   169

Conducting an ANCOVA in SPSS   170
Reporting an ANCOVA   174
ANCOVA Writing Hints   175
ANCOVA Exercises   175

12. Multivariate Analysis of Covariance (MANCOVA)   177
    Use of the Multivariate Analysis of Covariance (MANCOVA)   177
    MANCOVA Assumptions   179
    MANCOVA Research Design   180
    The Null Hypotheses for the MANCOVA   182
    Setting Up a Database for a MANCOVA   183
    Conducting a MANCOVA in SPSS   184
    Reporting a MANCOVA   190
    MANCOVA Writing Hints   190
    MANCOVA Exercises   190

13. Regression Analysis   193
    Use of Regression Analysis   193
    Regression Analysis Assumptions   196
    Understanding Some of the Math Used in Regression Analysis   197
    Null Hypothesis of the Regression Analysis   198
    Regression Analysis Research Design   198
    Understanding All of the Symbols   199
    Setting Up a Database for a Multiple Regression   199
    Conducting a Regression Analysis in SPSS   200
    Reporting a Regression Analysis   204
    Regression-Analysis Writing Hints   204
    Regression-Analysis Exercises   205

14. Data Reduction: Factor and Principal Component Analysis   207
    Use of Data Reduction Techniques   207
    Data Reduction Terminology   209
    Clarifying the Uses of Data Reduction   210
    Potential Limitation of Data Reduction   211
    Assumptions of Data Reduction   211
    Research Design and Null Hypothesis of Data Reduction   212

Setting Up a Database for Data Reduction 212
Conducting a Data Reduction Technique in SPSS 212
Reporting a Data Reduction Technique 219
Data Reduction Writing Hints 221
Data Reduction Exercises 221

15. Discriminant Analysis 225
    Use of Discriminant Analysis 225
    Model Development in Discriminant Analysis 225
    The Assumptions of Discriminant Analysis 228
    Research Design and Null Hypothesis
       of Discriminant Analysis 228
    Setting Up a Database for a Discriminant
       Analysis 229
    Conducting a Discriminant Analysis in SPSS 230
    Reporting a Discriminant Analysis 233
    Discriminant Analysis Writing Hints 234
    Discriminant Analysis Exercises 235

SECTION III | Reliability Analysis 237

16. Cronbach's Alpha 239
    Use of Reliability Analysis 239
    Assumptions of Reliability Analyses 240
    Cronbach's Alpha 240
    Setting Up a Database to Compute Cronbach's
       Alpha 241
    Conducting Cronbach's Alpha in SPSS 241
    Reporting Cronbach's Alpha 244
    Cronbach's Alpha Writing Hints 245
    Cronbach's Alpha Exercises 245

17. Split-Half Reliability 247
    Use of Split-Half Reliability 247
    Assumptions of Reliability Analyses 247
    Setting Up a Database for a Split-Half Reliability
       Analysis 249
    Conducting a Split-Half Reliability Analysis in
       SPSS 249
    Reporting a Split-Half Reliability Analysis 253
    Split-Half Reliability-Analysis Writing Hints 253
    Split-Half Reliability Exercises 254

SECTION IV | Nonparametric Tests 255

18. Chi-Square 257
    Use of Chi-Square Tests 257
    Assumptions of Chi-Square Tests 258
    Research Design and the Chi-Square Test 259
    The Null Hypothesis for the Chi-Square Test 259
    Setting Up a Database for the Chi-Square Test 259
    Conducting a Chi-Square Test in SPSS 259
    Reporting a Chi-Square Test 266
    Chi-Square Writing Hints 266
    Chi-Square Exercises 267

19. Mann-Whitney U Test 269
    Use of the Mann-Whitney U Test 269
    Assumptions of the Mann-Whitney U Test 270
    Research Design and the Mann-Whitney U Test 271
    The Null Hypothesis for Mann-Whitney U Test 272
    Setting Up a Database for a Mann-Whitney U Test 272
    Conducting the Mann-Whitney U Test in SPSS 273
    Reporting a Mann-Whitney U Test 276
    Mann-Whitney U Test Writing Hints 276
    Mann Whitney U Test Exercises 277

20. Kruskal Wallis H Test 279
    Use of the Kruskal Wallis H Test 279
    Post Hoc Tests for the Kruskal Wallis H Test 280
    Assumptions of the Kruskal Wallis H Test 280
    Research Design and the Kruskal Wallis Test 282
    The Null Hypothesis for the Kruskal Wallis H Test 284
    Setting Up a Database for a Kruskal Wallis H Test 285
    Conducting a Kruskal Wallis H Test in SPSS 285
    Reporting a Kruskal Wallis H Test 288
    Kruskal Wallis H Test Writing Hints 288
    Kruskal Wallis H Test Exercises 289

21. Spearman Correlation 291
    Use of Spearman Correlation 291
    Assumptions of Spearman Correlation 292
    Research Design and Spearman Correlation 293
    The Null Hypotheses for the Spearman Correlation 294

Setting Up a Database for a Spearman
   Correlation   294
Conducting a Spearman Correlation in SPSS   295
Reporting a Spearman Correlation   296
Spearman Correlation Writing Hints   296
Spearman Correlation Exercises   297

22. Wilcoxon Test   299

Use of the Wilcoxon Test   299
Assumptions of the Wilcoxon Test   300
Research Design and Wilcoxon Test   301
The Null Hypothesis for a Wilcoxon Test   301
Setting Up a Database for a Wilcoxon Test   302
Conducting a Wilcoxon Test in SPSS   303
Reporting a Wilcoxon Test   305
Wilcoxon Writing Hints   305
Wilcoxon Test Exercises   306

23. Friedman's Test   307

Use of Friedman's Test   307
Assumptions of Friedman's Test   308
Research Design and Friedman's Test   308
The Null Hypotheses for Friedman's Test   310
Setting Up a Database for a Friedman's Test   311
Conducting a Friedman's Test in SPSS   311
Reporting a Friedman's Test   313
Friedman's Test Writing Hints   314
Friedman's Test Exercises   315

Appendix   317
References   321
Index   325

PREFACE

Introductory Thoughts

There are few courses or subjects that inspire more trepidation in music education students (and, at times, teachers) than those which focus on the use of statistical procedures. This response can be for a multitude of reasons, including negative previous experiences, a fear of mathematics, or a philosophical belief that the art of music should not be intruded upon by mathematical analysis. In this book, I endeavor to explain the process of using a range of statistical analyses from inception to research design to data entry to final analysis by using understandable descriptions and examples from extant music education research. Through this methodical explanation, I hope to mitigate, if not alleviate, some of the apprehension experienced by so many students who are often forced to fumble through this experience by seeking assistance from outside the world of music education and with too little guidance.

In his text, *Introduction to Mathematical Philosophy*, Bertrand Russell claimed that the line between logic and mathematics is so blurred that the two forms of thinking are really one. He stated that

> [t]hey differ as a boy and man: logic is the youth of mathematics and mathematics is the manhood of logic. This view is resented by logicians who, having spent their time in the study of classical texts, are incapable of following a piece of symbolic reasoning, and by mathematicians who have learnt a technique without troubling to inquire into its meaning or justification.

Although his use of a gender-specific example and his seemingly placing a greater value on mathematics as the more mature endeavor may cause a reasonable contemporary reader to bristle, his fundamental point remains. In short, it is not desirable to distinguish between the mathematics of statistical procedures and the actual meaning of the findings or logic behind their computation and interpretation. Our profession would benefit from improving our understanding of statistics for everyone so that we avoid falling into one of the two categories outlined by Russell: those who understand the math without thought to context and meaning, and those who focus on content and meaning while avoiding symbolic reasoning.

Introduction to the Book

My intention for this book is that it be used by faculty and students in music education programs as a way of advancing the understanding of how parametric statistics are employed and interpreted in the social science field of music education. More specifically, I hope that this text will help researchers, teachers of research, readers of research, and students gain a better understanding of four aspects of music education research:

1. *Understanding* the logical concepts of various statistical procedures and what the outcomes mean in their specific contexts
2. *Critiquing* the use of different statistical procedures in extant research and developing research to ensure accuracy of use and interpretation of the results
3. *Applying* the correct statistical model for not only any given dataset, but also the correct logic determining which model to employ
4. *Reporting* the results of a given statistical procedure in a way that is clear and provides adequate information for the reader to determine if the data analysis is accurate and interpretable

The Aim of Statistics in Music Education Research

Two diametrically opposing views seem to have taken root in society in regard to quantitative measures used to examine our world. The first, often unfounded, view is often referred to as "big" data as being one of the more easily constructed and potentially understood basis for an argument or propellant for change. The second, often unfounded, view is an ever-increasing unease with the use of quantitative measures to understand phenomena as being over-simplistic or lacking in any nuance or meaning. Because these two contrary views seem to permeate discussions on this topic, clearly the need to discuss what statistics can and cannot accomplish and how the information obtained from such work can inform our practice as music teachers and learners is great.

In the world of music education, which is subject to fears of overly objectifying that which some would contend is a purely subjective art and that which happens within a social context, creating greater complexity than is often considered in quantitative research design, it is imperative to understand that the use of quantitative measures and research methodology is not a means to objectify the artistry in music or ignore the contextual nature of humans music making and learning. It is, rather, a means of organizing how we observe and analyze facets of musicianship and pedagogy in order to improve both.

In essence, the ultimate goal of all statistics is to help us to begin to think logically and analytically in order to reach informed conclusions based on the best available data. Otherwise, we are fated to perpetuate willful ignorance and forced to rely upon dogmatic thinking and practice. More specifically, we can use statistical procedures to determine, by using a sample of participants rather than the entire population while controlling for as many variables as possible, whether or not a stated hypothesis should be rejected or accepted. That is, can we as researchers and interpreters of research that employs inferential statistics generalize the findings of a sample of participants to the overall population? If a music education researcher is interested in examining the effectiveness of a new teaching strategy in comparison to the more established teaching strategy, the researcher will most likely not be able to study every single music teacher who would be likely to employ the new strategy. Therefore, she would find a sample, or small group that represents or is similar to the population, to study. How well that group represents the overall population or how well the researcher controlled for differing variables will inform whether or not we can make accurate generalizations.

Reading Formulas

Throughout this text, which is not formula heavy, I present the reader with the formulas for several statistical procedures. This effort is to help the reader understand the logic behind the tests and to begin to grasp the symbolic reasoning behind statistical testing. Moreover, a formula is really a summary of how to answer a question. Without a formula, we would need to start over each time we wanted an answer to a question. Learning to read a formula and understand the symbolic reasoning and algorithms that exist in it can be aided by thinking differently about and distinguishing between mathematics and arithmetic. Mathematics is the application of logic to study relationships, while arithmetic is the use of numbers to calculate (i.e., adding, subtracting, multiplying, and dividing). The best description I have read to distinguish the two practices is that arithmetic is to mathematics what spelling is to writing. In order to read a formula with meaning requires both arithmetic and mathematics, if you focus on one at a time, the task can be easier until you get more comfortable with reading them.

It can be most helpful when you are considering the arithmetic side to remember the order of operations. Remember to compute parenthesis data first, and then any exponents, followed by any multiplications or divisions, finally followed by any addition or subtraction. Generally speaking, the order of operations is

1. parenthetical computations (complete what is within parentheses)
2. exponents (e.g., squared numbers)
3. multiplication
4. division
5. addition
6. subtraction

When starting to focus on the mathematical side of any formula, remember that mathematics is about relationships. Addition is really the creation of a set from subsets that are not mutually exclusive (i.e., of like information). Subtraction is really about finding the difference between two sets (i.e., the difference between 8 and 3 is 5). Multiplication and division are opposite sides of the same coin and are often about finding relationships on different scales. You will find that in many formulas in this text a paired relationship between division and multiplication, as many formulas utilize squared integers and square roots at the same time in order to keep the formula balanced and on the same scale.

What might this balance look like? Let's start with a musical example before we examine the abstract form of the formula. A music education researcher and classroom teachers are often interested in what the average score on a test is in their studies or classrooms. For example, a high school choral-music educator (and his administrators) may be interested in what the average all-state choir audition score was for his students. To calculate this, the director would

1. sum up all of his students' audition scores
2. divide that number by the number of students who auditioned for all-state choir

Let's imagine that five students auditioned for all-state choir and received scores of 88, 98, 75, 81, and 79. To calculate the mean of these scores, you need only follow the steps above:

1: $88 + 98 + 75 + 81 + 79 = 421$
2: $421 \div 5 = 84.2$

Therefore, this director's students scored a mean of 84.2 on their audition scores. Without really knowing it, this director has just employed an algorithm to find the answer to his question. An algorithm is merely a set of step-by-step directions used to solve a problem. This algorithm can be written in a more concise way than the list of two directions. It would look more like a formula:

$$\text{average all-state audition score} = \frac{\text{sum of all student audtion scores}}{\text{the number of students who auditioned}}$$

In this representation we see the same algorithm described above in a new way. We see that the average all-state audition score equals the sum of all student audition scores divided by the number of students who auditioned.

Then we would add the data to the formula:

$$\text{average all-state audition score} = \frac{88 + 98 + 75 + 81 + 79}{5}$$

We could then solve the arithmetic:

$$84.2 = \frac{421}{5}$$

The goal for the learner would be to be able to apply this logic or algorithm in any context with any data. How could we take this logic and be able to apply it to any given new context? We would think of it as the algorithm or formula itself using abstract placeholders for the data we would enter. The abstract placeholders usually employed are the Greek symbols used by mathematicians. In our example of trying to find the mean or average, we could begin with a basic formula:

$$\bar{x} = \frac{\Sigma x}{n}$$

This formula is a visual representation of an algorithm that we all have calculated numerous times in our lives. We usually think through this calculation so quickly that we do not consciously think of the steps taken. Looking at the notational formula of this common statistical procedure can be intimidating for those not accustomed to such notation (much as our students can be intimidated by some musical notation). This equation includes $\bar{x}$ (pronounced X bar), which simply indicates a mean as well as the sum (Σ) of each individual score (x) divided by the number of scores included (n). As with learning to read standard musical notation, it takes some time to learn the symbolic meanings of the numeric and Greek characters. Once you have learned and practiced them, using such symbolic reasoning tools can help you streamline your reasoning and logic when thinking through statistical procedures (including the most simple to most complex statistical procedures), much as musical notation is a way for musicians to share musical ideas efficiently with those who have the same notational knowledge from something as simple as a three-note melody all the way to a fully orchestrated symphonic score and beyond.

ABOUT THE COMPANION WEBSITE

www.oup.com/us/statisticsinmusiceducationresearch

Oxford has created a website to accompany *Statistics in Music Education Research*. Material that cannot be made available in the book, namely, the data files that accompany each chapter, are provided here. The reader is encouraged to consult this resource in conjunction with each chapter, starting with Chapter 2. When these data files are most helpful are indicated in the text with Oxford's symbol ▶.

I | Introduction

1 | Fundamental Principles

Hypothesis Testing

Hypothesis testing is one of the primary practices in quantitative music education research. Basically, hypothesis testing is a means to make a rational (i.e., logical) decision based on the observed data. One can and should argue the nature of rationality, as it can take multiple forms and have fundamental ontological and axiological implications. However, for our purposes, and for the purpose of music education research, it is compelling to argue that rationality in any educational research endeavor is characterized by making the decisions that will most likely lead to student success.

To help keep a thread going *in this section on hypothesis testing*, we use a running example; an orchestra director has recently experienced young violin students bringing in new bows with colorful hair, both green and blue. The orchestra director has become curious if any difference exists in students' tone production on the basis of the color of the bow hair. The reason that our orchestra director wants to know if there is a tone production difference between the students with blue and green bow hair is to be able to help students create the best sound (yes, that in itself is interpretable) possible by using the hair that will lead to the greatest success for her and others' students.

To test a hypothesis, we use data (for the purposes of this text, quantitative data) to see if we should reject or accept the null hypothesis. For example, let us continue to consider our example of an orchestra director who is curious if a difference in the sound production of violin students who use bows with blue hair and students who play by using a bow with green hair exists. In order to find out if any difference does exist, she will need to test that hypothesis in some way, hence hypothesis testing.

A hypothesis is a proposed solution to a problem. As stated above, the rational outcome of testing a hypothesis is to ensure the greatest probability of success. To test a hypothesis we use data (for the purposes of this text, quantitative data) to decide if we should reject or accept the null hypothesis. The *null hypothesis* is the hypothesis in which a researcher assumes that no differences exist between any groups or sets of data. In our example, the null hypothesis would be that no difference exists between the students who use blue and green bow hair, or, stated differently, the mean of the first group will be effectively the same as the mean of the second group. When the evidence we find indicates that no difference between the two types of bow hair exists, we accept the null hypothesis that no differences exist. If, however, we find that the data indicates that a difference between the two groups most likely exists, we must reject the null hypothesis and concede that some difference likely exists.

Stating Hypotheses

It can be very helpful for a researcher to overtly state his hypothesis. Stating a hypothesis can give the researcher clarity in what he is trying to uncover and what method may be best able to do so, as well as give the ultimate reader of the research the best understanding of how to interpret and employ the information in her own practice or research. In a traditional formula, a null hypothesis (H_0) is stated

$$H_0: \mu_1 = \mu_2$$

where H_0 stands for the null hypothesis or hypothesis zero,
μ_1 stands for the mean of the first group, and
μ_2 stands for the mean of the second group

In our running example, the null hypothesis would indicate that the researcher assumes that the average of the score for the students with blue-haired bows will be no different from the average of the tone scores for students with green hair. Therefore, the null hypothesis of our running example would look like

$$H_0: \mu_{\text{blue bow hair}} = \mu_{\text{green bow hair}}$$

This is not to say that there can be only two groups. A stated hypothesis can, theoretically, continue unabated:

$$H_0: \mu_1 = \mu_2 = \mu_3 = \mu_5 = \mu_6 \text{ etc.}$$

Running Example: This might be the case should the research assume that the scores on a tone production test would be the same for five different groups of students, those with green bow hair, blue bow hair, red bow hair, yellow bow hair, and purple bow hair.

This musical-example null hypothesis could be notated as

$$H_0: \mu_{\text{green bow hair}} = \mu_{\text{blue bow hair}} = \mu_{\text{red bow hair}} = \mu_{\text{yellow bow hair}} = \mu_{\text{purple bow hair}}$$

Also, to be specific, the symbol mu (μ) represents the mean of the population rather than sample. If you were stating a hypothesis for the sample itself (rather then trying to generalize to the population), you would state it by using a capital M or X bar ($\bar{X}$):

$H_0: M_1 = M_2$ or, for our running example, $H_0: M_{\text{green bow hair}} = M_{\text{blue bow hair}}$

Much as with the formula discussed in the preface, music education researchers can use this abstract formula, or formulas similar to it, to describe the null hypothesis of their experimental studies. For example, if a researcher is interested in finding if a difference exists between young children who have learned to dance with folk music and those who only sing folk music, in terms of both groups' ability to find a pulse in music, researchers may state their null hypothesis as

$H_0: M_{\text{dancing folk students}} = M_{\text{singing folk students}}$

Alternative Hypotheses

Do researchers state only null hypotheses? No. If it makes sense to offer multiple hypotheses, that if the researcher is relatively sure (on the basis of previous research or evidence) that he will find differences in the means, then stating an *alternative hypothesis* may be warranted. If, for instance, our orchestra director felt sure that a difference would be found between students with green bow hair and students with blue bow hair, *but was unsure of the direction the difference would have taken*, she could state an alternative hypothesis. She could have stated her hypothesis As

$H_1 \text{ or } H_a = \mu_{\text{green bow hair}} \neq \mu_{\text{blue bow hair}}$

The abstract version of this hypothesis could be stated

$H_1 \text{ or } H_a = \mu_1 \neq \mu_2$ or...

H_1 and H_a are simply two different ways of indicating that a stated hypothesis is not the null hypothesis. The 1 indicates that it is different from the subscript 0 in the null hypothesis, while other authors employ the "a" as a subscript to indicate that it is an **a**lternative hypothesis.

Directional Hypotheses

What if the researcher believes that she will find a particular direction in her findings that is based upon previous research or theoretical framework? If this is the case, the researcher could state a directional hypothesis. When we return to our running example, if the researcher had sufficient evidence to believe that students with blue bow hair would have a much better tone production that students with green bow hair would, she may state her hypothesis in this manner. She may state a directional hypothesis—here indicated by using H_2 to distinguish it from the null hypothesis (H_0) or the alternative hypothesis (H_1 or H_a)—that indicates this assumed outcome:

$$H_2 = \mu_{\text{blue bow hair}} > \mu_{\text{green bow hair}}$$

In this stated hypothesis, we see that the researcher believes that the students with blue bow hair will have a greater outcome than students with green bow hair would. Therefore, in order to accept this directional hypothesis, the researcher would have to find strong empirical evidence that students with blue bow hair have better sound production than those with green bow hair. Otherwise, the researcher would rationally need to reject the hypothesis. The abstract version of this hypothesis could be stated

$$H_2 = \mu_1 > \mu_2$$

N.B. Be aware, however, that if you elect to utilize a directional hypothesis you should employ a slightly different type of statistical procedure. However, as the majority of this text describes methods used when you are assuming a null hypothesis (the more common practice), you should seek out additional information on such statistical tests.

Reporting the Hypothesis

How do we state hypotheses in research reports? In truth, we rarely do these days. Some music education researchers continue to state null hypotheses (see Example 1.1). This practice can make it easier to directly connect the results

---

EXAMPLE 1.1 STATING NULL HYPOTHESES IN PROSE FORM

In the current study, we sought to investigate the effects of recorded ensemble models on middle/junior high school and high school band students' performance self-evaluations, achievement, and attitude. Our null hypotheses were that (1) ensembles would demonstrate no difference in performance achievement between selections studied with or without a recorded model; (2) students' evaluations of their individual performances versus their ensemble's performance would be no different between the two conditions; (3) students' free-response self-evaluations would reveal no difference in attention to musical elements and to group (versus individual) performance issues between the two conditions; and (4) students would demonstrate no difference in attitude toward the model and no-model selections.

Morrison, S. J., Montemayor, M., & Wiltshire, E. S. (2004). The effect of a recorded model on band students' performance self-evaluations, achievement, and attitude. *Journal of Research in Music Education, 52*(2), 116–129.

EXAMPLE 1.2 USE OF RESEARCH QUESTIONS IN PLACE OF STATING HYPOTHESES

Three main research questions were addressed in this study: (1) What types of school district frameworks and classroom contexts are secondary music teachers operating within as they assess learning and grade students? (2) Which specific assessment and grading practices are most commonly employed by secondary music teachers? (3) Do any contextual or individual difference variables influence secondary music teachers' assessment and grading practices?

Russell, J. A., & Austin, J. R. (2010). Assessment practices of secondary music teachers. *Journal of Research in Music Education, 58*(1), 37–54.

---

of the study to the original statement of the problems and are usually stated in prose rather than a formula.

The more common practice is to state research questions rather than hypotheses. By using research questions, authors can incorporate several hypotheses within a single research question (see Example 1.2).

Had the authors stated the third research question in Example 1.2 as traditional null hypothesis (a hypothesis that assumes that there will be no difference between groups or examined variables) it could have been stated thus:

> *No contextual or individual difference variables influenced secondary music teachers' assessment or grading practices.*

As you can see, this one research question actually contains several possible null hypotheses. Russell and Austin identified two teaching context variables, school level (middle and high school) and teaching specialization (instrumental and choral). The null hypotheses may have been stated as

$H_0: \mu_{instrumental} = \mu_{choral}$
$H_0: \mu_{middle\,school} = \mu_{high\,school}$

In sum, music education researchers usually focus on one of three different types of hypotheses:

- Null hypothesis—no differences exist between any of the groups or sets of data being examined.
- Alternative hypothesis—a difference does exist between groups or sets of data, but the researcher has no inclination as to the direction of the difference.

- Directional hypothesis—a difference does exist between groups or sets of data and the researcher has a theory, based on evidence, of which direction the difference will be.

Measures of Central Tendency

There are three major measures of central tendency of any dataset: mean, median, and mode.

- Mean—the average of all of each of the individual data within a set.
- Median—the point at which 50% of the cases in a distribution fall below and 50% fall above.
- Mode—the most commonly occurring number within a set.

For example, consider a music education researcher who is interested in how well middle school band students achieved on a teacher-created sight-reading assessment. Ten students in a band class completed the assessment in which the possible score ranged from 0 to 10. Students in the class scored thus:

TABLE 1.1 Example Student Assessment Scores

STUDENT	SCORE
1	8
2	8
3	7
4	8
5	9
6	3
7	8
8	6
9	9
10	10

From the preface of this text, we know the formula for computing the mean:

$$\bar{X} = \frac{\Sigma x}{n}.$$

The mean of this dataset is 7.6. The median for this set of scores is 8. We can find this by first lining the scores up in order:

$$3, 6, 7, 8, \underline{8, 8}, 8, 9, 9, 10$$

We can see that by finding the number in the middle of the scores we find an 8. In this example, we are aided by the multiple 8s. However, with an even quantity of numbers in the set, the median is less clear. Nonetheless, whether we select the fifth or sixth number as our median, it remains an 8. The other way to find the median when an equal number of scores exist is to find the middle two numbers (in this case, the fifth and sixth numbers) and then find the average of that number $(8+8=16 \div 2=8)$.

The mode for this example is also 8. This quantity is easily identified, as it is the most commonly occurring number in the dataset. Had these two numbers been different, we would have needed to find the mean between the two numbers to find the median.

Upon first glance it may seem that not all three pieces of information would be necessary. However, it is possible to mislead or be misled by focusing on only one of these three data points. A dataset could be bimodal (or trimodal and so on) and could have peaks or modes well away from the mean. Similarly, a median can be misleading if the dataset is skewed in one direction or another. Generally, all three pieces of information should be somewhat similar in a normally distributed dataset. For example, let's assume that our researcher gave the same sight-reading assessment to 10 other middle school band students from another school. These students scored thus:

TABLE 1.2 Additional Student Assessment Scores

STUDENT	SCORE
1	9
2	8
3	7
4	3
5	9
6	3
7	2
8	4
9	9
10	3

The mean of this dataset is 5.7. The median is 5.5. We calculated these two quantities by lining up the numbers in order as before. We see that the two numbers in the middle are 4 and 7. We find the mean of those two numbers is 5.5 $(4+7=11 \div 2=5.5)$:

2, 3, 3, 3, **4, 7**, 8, 9, 9, 9

So far, this calculation seems pretty good. At first glance, the mean and median seem rather close to each other. It appears that our middle school band students have a logical grouping in the classroom as far as sight-reading skill goes. However, if we look at the mode, we see that this is a bimodal set of data. The two modes are 3 and 9. Neither of these two numbers is relatively close to either the mean or the median. What does this mean? It means that the set of data is a little more complex than just knowing one version of central tendency can describe. Often, multimodal datasets indicate that some external force is influencing outcomes. For example, students in this band class may be mismatched on the basis of previous sight-reading instruction or experience. One set of students may have found the sight-reading material previous to the assessment and practiced the material. One group of students may have been tested on a day where additional variables influenced the outcome and so on. It becomes the role of the researcher to try to understand and explain why such phenomena exist in his data.

Normal Curve (Gaussian Distribution)

A normal distribution of the data is one assumption of the statistical procedures used in this text as well as in music education research. In statistics, an assumption of a test is a required characteristic of the data before the test can be accurately used, much as learning to match pitch with your voice is a required characteristic if you are a student wishing to sing in an auditioned choir. A *normal distribution* of scores has three major characteristics:

- The curve is peaked at the mean and has only one mode.
- The curve is symmetrical on both sides of the mean, median, and mode.
- The curve has asymptotic tails. Asymptotic tails simply mean that the tails of both sides of the curve approach but do not touch the actual axis.

See Figure 1.1.

You may be better able to get a sense of what a normal curve looks like by examining the *skew* and *kurtosis* of a dataset.

- Kurtosis refers to the amount of peak found in a sample. The word kurtosis is derived from the Greek word "kurtos," which means curved or

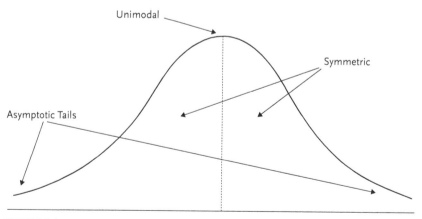

FIGURE 1.1

arching, so the use of the word in describing the contour of the data is apropos.
  o A sample that has a balanced kurtosis is known as *mesokurtic*. This usage makes sense, given the meaning of meso as the middle. A mesokurtic curve is one that is generally considered a normal curve.
  o A sample that has a slender dispersion and tight peak is known as *leptokurtic* (lepto means slender).
  o A sample that has a broad dispersion and flat peak is known as *platykurtic* (platy means broad; think of the flat, broad bill of a platypus).

See Figure 1.2.

We can imagine some musical examples that would lead to all of these kinds of distributions. A leptokurtic distribution may occur if a music education researcher used a research instrument eliciting responses that were too obvious and therefore participants in the study did not offer a great deal of variance in those responses. For example, asking in-service music educators in a survey how important they felt learning to sing or play music with others was. This question would most likely receive a rather leptokurtic response. A music education researcher may obtain data that is platykurtic if he uses a singular research instrument for a broad range of participants. Imagine using a single sight-reading test with high school band students in a school where there is only one period of a band that includes students who have been playing and reading for years and students who started playing their instrument that very year. You would expect the scores to be all over the place.

- Skew, or skewness is the quality of a distribution that defines the disproportionate frequency of certain scores.
  o A longer right tail than left corresponds to a smaller number of occurrences at the high end of the distribution; this is a positively skewed

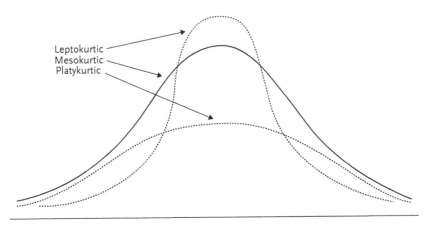

FIGURE 1.2

distribution. A shorter right tail than the left corresponds to a larger number of occurrences at the high end of the distribution; this is a negatively skewed distribution.

See Figure 1.3.

As stated above, this example is of a positively skewed distribution, as the longer tail is to the right. Generally speaking, *if the mean is greater than the median, the data is positively skewed. Conversely, if the mean is less than the median, the data is negatively skewed.*

The skewness of a normally distributed sample should be about zero. We will talk more about analyzing the normality of a dataset and about descriptive statistics in Chapter 2. It is possible to compute skewness without the assistance of the Statistical Package for the Social Sciences (SPSS) or any other software. Pearson (see Chapter 3, "Pearson Product Moment Correlation") created two different formulae for computing skewness. One uses the median of the sample; the other uses the mode. The formula for computing skewness with the median is

$$skewness_{median} = \frac{3(\bar{X} - Md)}{SD}$$

In this formula, X is the mean of the group, Md is the median and SD is the standard deviation (more to come about standard deviation in a few pages).

Similarly, the formula for computing skewness with the mode is

$$skewness_{mode} = \frac{\bar{X} - Mo}{SD}$$

As with the previous formula, X is the mean of the group, Mo is the mode, and s is the standard deviation. Keep in mind that samples with more than one mode will not be suitable for this particular formula.

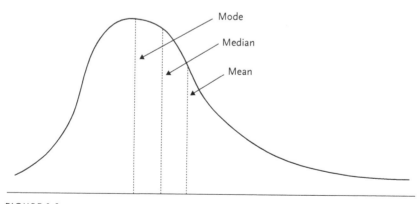

FIGURE 1.3

Why might a music education researcher obtain skewed data? Although data may be skewed for any number of reasons, an illustrative example might be helpful. One of the most common causes of skewed data is outlier data, that is, data points that do not look like the rest of the data. For example, a researcher may be interested in measuring how well students in her fifth-grade general music class can decode by using Curwen hand signs (a manual method of solfège). However, without the researcher's knowledge, the teacher from the previous year forgot to teach one of her 15 different classes of fourth-grade students to use hand signs at all. Those students, because of their different previous educational experiences, would most likely not look like the rest of the students on this measurement and would cause the data to be skewed.

A few other reasons that a music education researcher may obtain a non-normal distribution include the following:

- The data has been impacted by some additional process or force. In the example above, the students who did not learn Curwen hand signs impacted the set of scores by creating outlier scores. It was not that some students could not achieve; the data simply showed a flaw in the teaching process from the previous year.
- Not enough data has been collected. Generally speaking, the fewer the number of observations or participants in a study, the less likely the researcher is to find normally distributed data. If a high school music theory teacher has only five students in his class, it is not likely that he would find a normally distributed set of scores on their final Advanced Placement exams.
- The researcher used a poor or imprecise research instrument. If a researcher was interested in the ability of a group of high school orchestra students to phrase their all-county audition music by using a musical excerpt drawn in haste on the way into work that morning, it is likely that the data collection would not be accurate enough to obtain a normal distribution. The excerpt may work for the violins, but the extra string crossing for the violas may have obscured the real intent of the research findings.
- A non-normal distribution is the best descriptor of the data. Not all phenomena are best described by a normal distribution. Although a normal distribution is usually the goal in social science research, at times it is not the most accurate. Just as one example (and there are several, but they go well beyond the scope of this text), the Poisson distribution is a distribution that describes the likelihood of how many times an event would occur within a specific length of time. For example, a music education researcher may be interested in how many off-task behaviors are exhibited by percussion students during seventh-period band class. This data would be best described by using Poisson distribution rather than a normal Gaussian distribution.

We will examine a few ways to test for normality in Chapter 2.

Measures of Variability

Range

The most simple form of the measure of variability in any dataset is the range, which is simply the highest score achieved on a measure minus the lowest score on the measure. As we have already discussed, subtraction is really asking "what is the difference between?" So, range is asking what the difference is between the highest score and the lowest score. Using our data from the example above we can examine the range of scores as ten middle school band students completed a teacher-created sight-reading assessment and scored

3, 6, 7, 8, 8, 8, 8, 9, 9, 10

To find the range, we simply find the lowest score (3) and subtract it from the highest score (10). Therefore, we know that the range of the scores in this sight-reading test was 7.

Standard Deviation

Once we have a better grasp of where the central tendency in a dataset lies, the next thing we want to know is what the average deviation is from that point of central tendency (i.e., how well the data sticks together). Standard deviation is the average deviation from the mean. There are two types of standard deviation to consider, however; the standard deviation of an entire population and the standard deviation of the sample. The primary difference is that of the *degrees of freedom* (we will explores degrees of freedom a little later in this chapter).

The standard deviation of the population formula is

$$\sigma = \sqrt{\frac{\sum (x - \bar{x})^2}{N}}$$

where σ is the standard deviation of the entire population, $\sum$ is the sum, x is each value of the population, $\bar{x}$ is the mean of the values, and N is the number of values in the population. Let's say, however, that we do not know the entire value of x for the entire population. Then the formula needs more freedom (see degrees of freedom below). The formula for the standard deviation of a sample is

$$s = \sqrt{\frac{\sum (x - \bar{x})^2}{N - 1}}$$

where s is the standard deviation of the sample (rather than population), $\sum$ is the sum, x is each value of the sample, $\bar{x}$ is the mean of the sample values, and N is the number of values in the population.

Here we can take a look at a musical example. Let's consider the work of a music education researcher who was interested in examining the impact of marching-band participation on concert-band students' tone production. In this process, the researcher rated the participants' tone by using a researcher-created tone test between 1 (non characteristic) and 5 (very characteristic). Here is the data she obtained:

- Ten students took the test.
- The scores were 3, 5, 4, 5, 3, 3, 4, 5, 2, 5.

With this information, we can begin to compute the standard deviation. As this is just a sample and not *all* high school band students, we use the formula for the sample and not the population:

$$s = \sqrt{\frac{\sum(x-\bar{x})^2}{N-1}}$$

1. The first step is to find the mean of all the scores. We know how to do so (see the preface of this text). The mean of the 10 scores is 3.9

$$s = \sqrt{\frac{\sum(x-3.9)^2}{10-1}}$$

2. Then we take each individual score and subtract the mean and then square that answer. Remember that subtraction is really asking the question "what is the difference between?" So, you are looking for the difference between the mean and each student's score. The squaring (or multiplying), as discussed in the preface, is really about finding relationships in different scales. This quest is also why we must later find the square root in order to bring the final answer back into the same scale as the raw data.

TABLE 1.3 Solving for the Standard Deviation

STUDENT SCORE	SCORE MINUS THE MEAN	SCORE MINUS THE MEAN, THEN SQUARED
3	−.09	.81
5	1.1	1.21
4	0.1	.01
5	1.1	1.21
3	−.09	.81
3	−.09	.81
4	0.1	.01
5	1.1	1.21
2	−1.9	3.61
5	1.1	1.21

3. Next, we sum (Σ) the squared scores. We find that the sum of the squared student scores is 10.9.

$$s = \sqrt{\frac{10.9}{10-1}}$$

4. We can now easily compute the denominator of the formula:

$$s = \sqrt{\frac{10.9}{9}}$$

5. We can now easily solve the fraction:

$$s = \sqrt{1.21}$$

6. The last step is to find the square root of the remaining number to find the standard deviation:

$$s = 1.10$$

What does this score mean? Generally, if the scores are normally distributed, it means that roughly two-thirds of all of the scores on the test will most likely be within 2.8 (mean minus the standard deviation and 4.0 (mean plus the standard deviation), showing the reader the dispersion of scores or variance.

Variance

Variability is the amount of spread or dispersion in a set of scores, while variance is the square of the standard deviation, and another measure of a distribution's spread or dispersion. Generally speaking, the smaller the standard deviation and variance, the more agreement existed in responses to the stimulus or responses or scores were more similar between all of the participants in the study.

$$\text{variance} = \sigma^2 \text{ (if variance of the population) or } s^2 \text{ (if variance of the sample)}$$

In fact, if you have computed the standard deviation of a score, you have already computed the variance of the same score. The variance is the standard deviation in step 5 above, prior to your finding the square root of the final number. So in our musical example above, the variance of the sample was 1.21, or 1.10^2.

Why is it important for music education researchers to know about the variance in their data? It is important because without knowing this information you cannot compare different groups effectively or accurately. You would not know the most appropriate test to use, as discussed in the sections "Parametric Statistical Procedures" and "Nonparametric Statistical Procedures.

N.B. See the section below regarding degrees of freedom for an additional example of standard deviation solved by using our running example.

For example, if an orchestra director gave her students a tone-production test and the standard deviation was 0 (and therefore the variance as well; remember variance is just the standard deviation squared and 0 squared is 0), every single student earned the exact same score on the test. If there the possible range of the scores was 0–100 and the standard deviation was 8.14 (variance = 66.26), a greater amount of variance existed in scores. Similarly, if a researcher asked a question on a survey of university music education professors regarding their philosophy of assessing musical learning and one survey item had a standard deviation of .04 and another had a standard deviation of 4.3 (both on a 5-point scale), the researcher could state that the participants showed greater agreement on the item with the .04 standard deviation, as a smaller amount of variance existed in participant responses.

What is the relationship between a normal distribution and standard deviation? If you add percentages, you will see that usually around 68% of the distribution of scores lies within one complete standard deviation of the mean (the peak of the curve). Around 95% of the distribution falls within two standard deviations of the mean, while almost all of the distribution (99.7%) falls within three standard deviations. When these distributions are found within a normal curve, it is often referred to as the *empirical rule*. See Figure 1.4.

For example, if a music teacher gave rhythm reading test to all 100 of her seventh-grade choir students and found a normally distributed set of scores with a mean score of 50 and a standard deviation of 10, she could see if the distribution of student rhythm-reading scores met the empirical rule. If it did, she would most likely find that 68 of the students scored within one standard deviation of the mean, or scored between 40 and 60 points, while about 27 other students scored either 30 to 40 points or 60 to 70 points on the test. The remaining five students most likely scored below 30 points or above 70 points on the test.

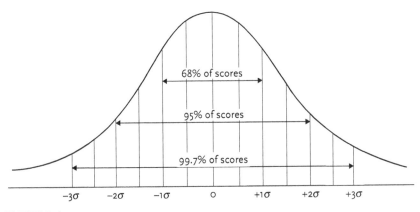

FIGURE 1.4

Error

Before a music education researcher can decide whether or not he wants to reject or accept the null hypothesis (or whatever type of hypothesis had been employed), the researcher needs to consider the error inherent in all social research designs such as those that music education researchers most commonly use There are two principle types of errors a researcher should consider:

Type I error—when a researcher rejects the null hypothesis when in fact the null hypothesis was accurate

Type II error—when a researcher accepts the null hypothesis when in fact the null hypothesis was not accurate

Researchers control for Type I error by setting the most appropriate alpha, or significance level. The most common alpha in social research is .05 (see the section titled "Significance" below for more information regarding alpha and statistical significance). However, if researchers want to strengthen their argument for rejecting the null hypothesis, they can set it lower (e.g., .01), which is also known as a more conservative alpha. A more conservative alpha makes it harder for a researcher to reject a null hypothesis (in other words, to find a significant difference) but may increase the threat of a Type I error. A more liberal alpha (e.g., .10) makes it easier to find significance, thus increasing the threat of a Type II error. Researchers can also adjust alpha on the basis of the number of comparisons being made in order to mitigate a Type I error. The most common version of this is known as a Bonferroni adjustment or correction in which alpha = .05 divided by the number of comparisons. The formula for the Bonferroni adjustment is

$$\alpha \div \text{number of comparisons}$$

For example, if a music education researcher was interested in collecting large amounts of data regarding student achievement in music as well as three of the major factors that may influence that achievement (i.e., parental support, teacher background, socioeconomic background), that researcher may want to try to avoid Type I error by employing a Bonferroni adjustment. In this case, the researcher would most likely be completing three analyses, one regarding parental support, one regarding teacher background, and one regarding socioeconomic background. To complete a Bonferroni adjustment, the researcher would take the standard social science alpha (.05) and divide that by 3 to obtain the new alpha of .02 (after rounding). Therefore, in order to reject any null hypothesis, the researcher would require an outcome on a test below .02.

EXAMPLE 1.3 MITIGATING TYPE I ERROR USING A BONFERRONI ADJUSTMENT

The conservative Bonferroni correction was used to attempt to control for the increased chance of Type I error that results from multiple comparisons. Accordingly, an alpha level of .001 was considered the threshold for statistical significance (i.e., alpha of .05/30 comparisons).

Miksza, P., Roeder, M., & Biggs, D. (2010). Surveying Colorado band directors' opinions of skills and characteristics important to successful music teaching. *Journal of Research in Music Education, 57*(4), 364–381.

As seen in Example 1.3, music education researchers explored what in-service band directors thought were the important skills for being a successful music educator. The researchers hoped to use this information to inform their practice as music teacher educators. However, because they collected so much data and needed to complete so many analyses, it was possible they might commit a Type I error. Therefore, like many social science researchers, these music education researchers employed the Bonferroni adjustment to mitigate a Type I error.

Music education researchers often control for Type II error by examining the power of the test outcome. Power is most commonly discussed in terms of having a sample that is large enough in relation to the overall population and by not making alpha so conservative that possible significant findings are missed. See Example 1.4.

Is there ever a clear answer? No. *All researchers who employ inferential statistics make Type I and Type II errors.* Our job is to understand the information provided as well as we can in order to make the best decision possible. It is preferable, however, to make a Type II error than a Type I error, especially if the study is not an exploratory one. For an illustration of this point, think about the impact a decision can have in a court of law. A Type I error is most akin to a guilty verdict when, in fact, the defendant was innocent. A Type II error is similar to finding a guilty person innocent. Obviously neither is ideal, but the greater social injustice is the innocent person being found guilty. Similarly, it is better to avoid basing your suggestions and implications on less powerful findings than missing some significant findings you may uncover at a later time as you develop your research agenda. If you want your suggestions as a researcher to have greater impact on the music classroom, make only the suggestions that you can support with strong empirical evidence.

> ### EXAMPLE 1.4 BALANCING TYPE I AND II ERROR
>
> Many statisticians subscribe to the strict use of the Bonferroni adjustment to protect against this inflation. However, the adjustment to lower the alpha level, while reducing Type I error, also escalates the possibility of Type II error. Therefore, to strike a balance between inflating the probability of Type I error with no adjustment and raising the potential for Type II error with a strict Bonferroni adjustment, a more stringent alpha level of $\alpha = .01$ was utilized for this study rather than the more traditional alpha level of $\alpha = .05$.
>
> ---
>
> Bright, J. (2006). Factors influencing outstanding band students' choice of music education as a career. *Contributions to Music Education, 33*(2), 73–88.

Types of Variables

Researchers often separate variables into two different types, independent variables and dependent variables. This sorting is especially helpful when researchers are trying to identify the correct statistical procedure to use. An independent variable is the treatment variable that is manipulated or selected by the researcher to see its impact on the dependent variable. The dependent variable is the outcome variable, or the predicted variable, in a regression equation. Think of it this way: dependent variables *depend* upon the independent variables. So, for example, if a researcher was interested in the impact of gender and high school grade point average (GPA) on a student's SAT score, the gender and high school GPA are independent variables and the SAT is the dependent variable.

For example, if a music education researcher was interested in the effect of the use of a computer-based practice software on middle school guitar students' musical memorization, one method to study it would be to ask one class of guitar students to use the software to help them memorize their music, while another class of students would strive to memorize their music without the aid of the software. Then the researcher would give a music memory test on the selected music to all of the students. In this example, the use of the software or not is the independent variable, while student scores on the music memorization test is the dependent variable.

Types of Data

It is important for a music education researcher or reader of music education research to understand the different types of data used in statistical procedures. The type of data being used informs the reader as to the most appropriate statistical procedure to use, as well as how to better interpret the data.

Nominal data—data that is *categorical,* such as gender, hair color, or what instrument a person plays. Usually reported in frequency tables.

Ordinal data—describes the order that data belongs to, but not necessarily the scale or interval that separates the data. The numbers you assign to this type of data represent the rank of the particular finding. Despite the common use of Likert-type scales as interval data, they are most accurately categorized as ordinal data (i.e., never, rarely, sometimes, often, always are really ordinal *categories* of data) because we do not really know what the difference between rarely and sometimes is, or if the difference between rarely and sometimes is the same difference as between sometimes and often.

A musical example of ordinal data might be students responding to a series of questions regarding the extent to which they enjoy the timbre of an instrument.

1 = not at all
2 = slightly
3 = moderately
4 = very
5 = extremely

Now, a reader can easily place these responses in order from least enjoyment to the most enjoyment of an instrument's timbre. However, neither the participant in the study nor the researcher really knows, for example, if the distance on the scale between slightly and moderately is the same distance as between moderately and very. So, this is ordinal data, which is not really on any type of intervallic scale.

Interval and ratio data—data on a *continuous* scale where differences between data points can be interpreted. The differences lie in whether or not the data has a natural zero point. Interval data has no natural zero point, while ratio data does. The next issue is helping readers to understand what a natural zero point is: a point in which 0 really means that none of that variable exists. Height, time playing the cello, and GPA are examples of scales that have a natural zero point. Another example may be a scale of practice time for high school students. In this instance, a 0 really means that no practicing has occurred and the difference between practicing for 10 minutes and 20 minutes is the same ratio as the difference in practicing for 40 minutes and 50 minutes. *In the end, the difference between interval and ratio data is important to understand, but not necessarily important to how the data is treated in most statistical analyses.*

In sum,

Nominal = variable attributes can only be named.
She plays cello, he plays viola.

Ordinal = variable attributes can be placed in order.
He enjoys the sound of cello more than the sound of the viola.

Interval and ratio = distance between variable attributes are meaningful.
She has played cello twice as long as he has played viola.

Generally speaking, it is very helpful to the researcher to be able to place data into two primary types of data, *categorical* and *continuous*. This sorting will have a great impact on the researcher's ability to select the most appropriate statistical procedure to use. For the purposes of this text and in general, categorical data is logically placed into the categorical column, while interval and ratio data are placed in the continuous column. Ordinal data is less clear. Many social science researchers treat ordinal data as continuous even though the logic of the scale is categorical. Generally, the outcomes of the statistics are relatively the same regardless of how the researcher treats the data. However, researchers are under even greater obligation to report how well the data met the assumptions of the statistical procedure used if they treat ordinal data as continuous rather than categorical.

Types of Statistical Procedures

Parametric Statistical Procedures

As much of this text is organized by distinguishing the difference between parametric and nonparametric statistics, it will help the reader to understand these differences early. Parametric statistics are used for the inference from a sample to a population. Parametric statistical procedures generally have more assumptions (i.e., requirements that mean the data is adequate for the test) than nonparametric tests. Generally speaking, *parametric statistical procedures require interval or ratio data*. The most common assumptions in parametric tests include the following:

1. *Observations are independent*—the observations used to examine a phenomenon are not acted on by any outside influence common to any of the observations, including one another. For example, if a researcher was interested in the performance achievement of two groups of violin students (those who have taken private lessons and those who have not), the observations made by the researcher of the playing skill of those students with private lessons are independent of the observations of the playing skill of the students without private lessons. In other words, the scores of students without private lessons will not go down or up because of the scores of the students who did take private lessons.
2. *Observations are taken from normally distributed samples or populations*—the samples or groups being examined have normally distributed scores around the mean. Otherwise, we might be making a Type II error (see above for a discussion on error as well as normal

distribution) because we may be comparing the outliers of one group to the normally distributed other group. As in our previous example, if the scores of the students who have taken private lessons fit in a normal bell curve and the scores of the students who have not taken private lessons are all over the place, it is too difficult to see if real differences would exist beyond chance. It would be like trying to compare apples to oranges. They may both be basically round and fruit, but completely different in taste and nutrition. If both groups' scores were all over the place, it would simply be impossible to distinguish one group from the other with any accuracy.

3. *The samples or populations must have relatively equal variances*— similar to the second assumption discussed above. If the variance between the compared groups is too difference, the groups do not look similar enough to compare with any accuracy. The threat of a Type I error would be too great.

Nonparametric Statistical Procedures

Nonparametric statistical procedures are those which are not dependent upon the same assumptions as parametric tests. These types of tests do have lesser statistical power (sensitivity or ability to correctly reject the null hypothesis) than the parametric tests. That is not to say, however, that they are not often the most appropriate and useful tests to employ. Nonparametric tests are often ignored by researchers when they should be used when the data do not meet the requisite assumptions for parametric tests. Using parametric tests when nonparametric tests should be used can lead researchers to find inaccurate results and increase the likelihood of committing a Type II error. Nonparametric tests have two common assumptions:

1. *Observations are independent*— the same assumption as discussed above for parametric tests.
2. *The variable has some underlying continuity or logic*—there is some logical reason for the variable to be used as a means for comparison.

The example we used above in regard to comparing the performance achievement of violin students who have taken private lessons with that of students without private lessons is an interesting one with which to discuss the idea of underlying logic. Obviously, it is quite common for music educators and music education researchers to see classes with both of these types of student experiences, and it can be assumed that the additional and personal instruction that comes with private lessons would impact performance achievement. Therefore, a researcher may wish to use this factor as a variable in a nonparametric test, given the logical continuity to the possible impact of the independent variable (lessons) on the dependent variable (performance achievement).

It is up to the researcher, through careful reporting, to make the argument that the logic exists to meet this assumption.

Significance

Researchers often claim that their findings are statistically significant. What does this claim actually mean? For much of social science research, we set an a priori significance level, or alpha (α), of .05. If the outcome of a statistical procedure (p value) is below .05 (i.e., .049 and lower), we claim that it is a significant finding. What this statement really means is that the finding was most likely not due to chance and that if we reject a null hypothesis (i.e., find a significant difference), it is likely that the same difference will hold true for the entire population and not just the examined sample.

N.B. Findings statistical significance is a binary phenomenon. That is, a finding is either statistically significant or not. There is no "approaching significance" or "trend" toward significance. These are phrases employed by researchers who are invested in the findings. Negative results (i.e., not significant results) are just as informative as significant results and should be reported as such without the need to suggest a possible different outcome.

D.M. (*dignum memoria*). Keep in mind that the a priori (before the fact) significance level set by the researcher is the alpha (α), while the p value is the probability value or the likelihood of the statistic occurring because of chance. Some research reports have incorrectly used these symbols interchangeably.

For example, a researcher examined the difference between the Music Aptitude Profile scores of incoming freshman music-education vocal students and incoming freshman music-education instrumental students and found a statistically significant difference in the scores of two groups; vocal music education majors obtained higher scores than instrumental music education students. So, if the study was conducted well, it would seem to hold true that *all other variables remaining the same*, incoming music-education vocal students *anywhere* would have better Music Aptitude Profile scores than incoming instrumental music education students *anywhere*.

Statistical significance is not the only versions of significance that a social science researcher should examine and report. An equally important notion of significance to that of statistical significance is that of *practical significance*. This concept is usually measured through some form of variance examination of effect size. These effect sizes are discussed within each chapter for each procedure. In music education research, it behooves us to help guide teachers and policy makers to the most impactful changes. This guidance can be informed by examining the practical significance of a statistical outcome. Because so many tests of significance can find statistically significant outcomes, it becomes the ethical job of a researcher to offer more information about the test so that the reader as well as the author can consider the practical significance of the findings. Researchers most commonly accomplish this task by reporting effect size (i.e., the magnitude or strength of any relationship).

For example, in a study of 1,200 high school choir students, a music education researcher found that, statistically, students who sang in choral organizations outside of school as well as in school were more likely to stay in school choir up to graduation. However, after examining the effect size, or magnitude, of this finding, the researcher discovered that this one variable does seem to explain quite a bit of the difference in tone production between the two groups. Therefore, the researcher could also claim practical significance. Had she not found a strong effect size, she might have been required to temper her statistically significant findings with a warning to her reader that her findings might have little practical meaning.

A third level of significance that is rarely discussed is that of *policy significance:* a type of significance that indicates the findings of a study should immediately inform policy. The researcher has found statistical significance and practical significance, and can explain such a large effect on the basis of the variables studied, that change should occur. Unfortunately, this level of significance, although informed by outcomes of a statistical procedure, is open to interpretation by the researcher and reader, and is rarely found in social science research. Excellent and thorough research reporting of impressive findings are the only means for the researcher to make an argument for policy significance.

For example, a researcher found a statistically significant difference between students in two different university oboe studios in regard to the impact of shaping their own reeds on tone production. After examining the effect size or magnitude of her findings, she saw that the variable of students shaping their own reeds or not seemed to explain such a large amount of the difference between the two groups that we should change policy and teach all university oboe students to shape their own reeds everywhere.

In sum,

- *Statistically significant*—interesting, but we most likely need more data.
- *Practically significant*—interesting; we should really look into this phenomenon.
- *Policy significant*—we should alter our current practices on the basis of this new information

Degrees of Freedom

The number of degrees of freedom is the number of values in the final calculation of a statistic that are free to vary. Another way to think of it is as a means of keeping score: a dataset contains a number of observations, say, n. The observations constitute n individual pieces of information. These pieces of information can be used to estimate either parameters or variability. In general, each item being estimated costs one degree of freedom. The remaining degrees of freedom are used to estimate variability.

> A single sample—there are n observations. There is one parameter (the mean) that needs to be estimated. That leaves $n - 1$ degrees of freedom for estimating variability.
>
> Two samples—there are $n_1 + n_2$ observations. There are two means to be estimated. That leaves $n_1 + n_2 - 2$ degrees of freedom for estimating variability.

Another way to think about degrees of freedom is to give flexibility to an outcome. For example, the formula for the standard deviation of a sample as we learned earlier in this chapter is

$$\sigma = \sqrt{\frac{\sum(x - \bar{x})^2}{N - 1}}$$

In this formula, $N - 1$ is the degrees of freedom. So if the sample had 10 people in it (like our example earlier in this chapter), we are now saying it had 9. What happens to the outcome of solving fractions when the denominator is smaller? The outcome increases. In this regard, because we have data only from a sample and not the entire population, we are stating overtly that because we do not have all the information, we are artificially creating a larger standard deviation because we are unable to be as exact as if we had had all the possible information. That is why the formula for the standard deviation of a population when you do have all of the information is

$$\sigma = \sqrt{\frac{\sum(x - \bar{x})^2}{N}}$$

We have all of the information, so we do not need to increase the variance. As you can see, we are not increasing the standard deviation. We know enough to be more exact.

Let's take a look at our actual musical standard-deviation example from earlier in the chapter and compare the version we have already worked out to a version with no degree of freedom. *The shaded row shows where the computational difference in the two processes begins.*

As you can see and as discussed above, the standard deviation of the population (everyone possible who fits the criteria) is made smaller

TABLE 1.4 Degrees of Freedom Compared to No Degrees of Freedom in Standard Deviation

WITH A DEGREE OF FREEDOM (SAMPLE)	WITHOUT A DEGREE OF FREEDOM (POPULATION)
$s = \sqrt{\dfrac{\sum(x-\bar{x})^2}{N-1}}$	$\sigma = \sqrt{\dfrac{\sum(x-\bar{x})^2}{N}}$
$s = \sqrt{\dfrac{\sum(x-3.9)^2}{10-1}}$	$\sigma = \sqrt{\dfrac{\sum(x-3.9)^2}{10}}$
$s = \sqrt{\dfrac{\sum(x-3.9)^2}{9}}$	$\sigma = \sqrt{\dfrac{\sum(x-3.9)^2}{10}}$
$s\sqrt{\dfrac{10.9}{9}}$	$\sigma\sqrt{\dfrac{10.9}{10}}$
$s = \sqrt{1.21}$ *	$\sigma = \sqrt{1.09}$ *
* this step ($s = \sqrt{1.21}$) also gives us the variance of the sample, as it is the standard deviation squared	* this step ($\sigma = \sqrt{1.09}$) also give us the variance of the population, as it is the standard deviation squared
Sample standard deviation: $s = 1.10$	Population standard deviation: $\sigma = 1.04$.

simply by changing the denominator of the fraction that represents the number of participants being accounted for in the logic, as we can be more exacting in our prediction of variance. With the sample standard deviation, more freedom (or, to be exact, more error) is included and thus variance is larger.

If we applied what we learned earlier about the normal curve and the data in this example found earlier in the chapter fit a normal distribution (actually quite challenging to achieve with only 10 data points), we could expect that roughly 68% of the students *in this sample* should fall between 2.8 and 5.0 ($n = 7$). This calculation is determined by examining the mean (3.9) and one standard deviation (1.1) on either side of the mean. Does this supposition hold true? In this instance, not surprisingly, it does not. Actually, all but one score (90%) falls within one standard deviation of the mean. One easy way to check is to look at a chart of the data: most likely too many of the scores are clustered at the mean, which would lead to a leptokurtic distribution. This researcher could either elect to employ nonparametric analyses or collect more data and examine whether or not

TABLE 1.5 Selecting the Correct Statistical Analysis for Your Research

STATISTICAL ANALYSES PARAMETRIC	STATISTICAL ANALYSES NONPARAMETRIC	INDEPENDENT VARIABLES		DEPENDENT VARIABLES		CONTROL VARIABLES/ COVARIATES
		NUMBER OF INDEPENDENT VARIABLES	DATA TYPE	NUMBER OF DEPENDENT VARIABLES	DATA TYPE	
	Chi square	1	Categorical	1	Categorical	0
One sample t-test		NA	NA	1	Known continuous	
Independent-samples t-test	Mann-Whitney U	1	Categorical (only 2 groups or levels)	1	Continuous	0
Dependent-samples t-test	Wilcoxon test	1	Categorical (only 2 groups or levels)	1 observed 2 times	Continuous	0
ANOVA	Kruskal-Wallis H test	1 or more	Categorical	1	Continuous	0
Repeated measures ANOVA	Friedman's test	1 or more	Categorical	1 observed 2 or more times	Continuous	
ANCOVA		1 or more	Categorical	1	Continuous	1+
MANOVA		1 or more	Categorical	2+	Continuous	0
MANCOVA		1 or more	categorical	2+	Continuous	1+
Pearson correlation	Spearman correlation	1	Dichotomous or continuous	1	continuous	0
Regression		1	Continuous	1	Continuous	
Multiple regression		2 or more	Continuous	1	Continuous	0
Factor analysis		Any number of categorical and/or continuous variables to be reduced				
Discriminant Analysis		2 or more	Categorical or continuous	1	Categorical	0

the subsequently larger dataset met the parametric assumption of normal distribution.

Moving Forward

Now that you have a grasp of some of the concepts that will form the basis of the coming chapters, I will outline how the rest of the text is organized. I have organized this text into four major sections and endeavored to sequence the text such that each chapter builds upon the learning in previous chapters. Thus, in the first section of the book, I discuss some principles and definitions that will help the reader to contextualize the information that follows throughout the text. I also discuss descriptive statistics, upon which all subsequent analyses are based, so that you can make connections between raw data and the more nuanced procedures discussed in the text.

In the second section of the text, I focus on common parametric statistical procedures, which are commonly employed in music education research. The third section, although short, is important in that it describes reliability measures in statistics, measures that can help a reader better interpret the appropriateness of a researcher's claim. In the fourth and final section, I focus on the less often employed yet integral nonparametric statistical procedures. Within each chapter, I organize reading similarly to increase ease of reading as well as comprehension. Each chapter uses examples adapted from previously published research within the field of music education and has two datasets for you to work through. The first dataset is the one that I use throughout each chapter to allow you to go through the process once with the text; the second is for you to try on your own by using the same methods as the first. Moreover, at the end of this text is an appendix with some of the analyses from the second dataset from each chapter. This section will help you know if you are on the right track with your analysis.

Although I have endeavored to make this text flow in a logical manner to be most helpful to the reader, some may wish to jump directly to a chapter to help them better understand either a research study they are reading or one that they are currently undertaking themselves. To help those who are trying to look for the chapter that would be most beneficial to their specific current work by identifying which statistical analyses would be most useful, here is a graph to help readers identify the most likely statistical procedure for their work.

It might also be helpful to think about the type of question you are trying to answer when selecting a statistical procedure:

TABLE 1.6 Questions to Be Answered by Each Statistical Analysis

PARAMETRIC TEST	NONPARAMETRIC TEST	QUESTION TO BE ANSWERED
	Chi square	Does a difference exist between groups?
One-sample t-test		Does a difference exist between the sampled group and a known mean?
Independent-samples t-test	Mann-Whitney U	Does a difference exist between two groups' means on one dependent variable?
Dependent-samples t-test	Wilcoxon test	Does a difference exist between two observations of the same phenomenon (paired samples) on one dependent variable?
ANOVA	Kruskal-Wallis H test	Does a difference exist between two or more groups on one dependent variable?
Repeated-measures ANOVA	Friedman's test	Does a difference exist between three or more observations of the same phenomenon (paired sample) on one dependent variable?
ANCOVA		Do differences exist between two or more groups after controlling for one or more covariate (control variable) on one dependent variable?
MANOVA		Do differences exist between two or more groups on two or more dependent variables?
MANCOVA		Do differences exist between two or more groups after controlling for one or more covariate on two or more dependent variables?
Pearson correlation	Spearman correlation	To what extent are two or more variables related to one another and is the relationship direct or indirect?
Regression		To what extent can the one independent variable predict the outcome of the dependent variables?
Multiple regression		To what extent can the two or more independent variables predict the outcome of the dependent variables?
Factor analysis		Given a large number of variables collected, what are the underlying structures or least number of latent variables (inferred from the data rather than overtly collected) that really exist?
Discriminant analysis		How well can the collected independent variables predict membership in a categorical dependent variable?

2 | Descriptive Statistics

Use of Descriptive Statistics

Descriptive statistics are the beginning building blocks upon which music education researchers can describe and analyze their data. There are two primary reasons that descriptive statistics should be used by music education researchers:

1. To report the characteristics of participants in order to help readers make decisions about the generalizability of the sample to the overall population or to their own particular context. See Example 2.1.
2. To report the raw scores on research instruments used in the study so that readers may better understand the appropriateness of any inferential statistical procedure used later in the study or as the primary means of analysis, given the inability to use inferential tests or the inappropriateness of using inferential procedures. See Example 2.2.

Many times, however, it is more prudent and efficient to report descriptive statistics in table format, because of the large amount of information, and to save space and to increase readability of the text as well as the readers' ability to easily compare the findings. See Figure 2.1.

As discussed in Chapter 1, two primary types of descriptive statistics used in music education research exist.

1. Indicators of central tendency (mean, median, mode)
2. Indicators of variance (standard deviation, standard error)

The most commonly employed indicator of central tendency is the mean. However, the mean does not always give us *all* the information we need. Two groups that have the same mean but have different variance (as often indicated by standard deviation) can be quite different groups. See Figure 2.2.

As you can see in Figure 2.2, although the two groups may have similar means, they do not share similar variance. The dashed-line group scores are

EXAMPLE 2.1 USING DESCRIPTIVE STATISTICS TO DESCRIBE PARTICIPANTS AND CONTEXT

Surveys were returned from principals representing various regions of the United States: Midwest (32%), Northeast (27%), South (26%), and West (15%). These proportions closely reflected the membership of the population from which the sample was drawn. One of the responses was not used because the principal no longer worked at the elementary level. Respondents ($N = 214$) reported the length of their service as elementary school administrators to be as follows: under 1 to under 5 years (28.5%), [1] 5 to under 10 years (30.8%), and 10 or more years (40.7%). Most principals worked in suburban (40.7%) and rural (39.3%) locations, with a smaller percentage in urban settings (20%). The majority of schools (92.5%) required music education, a handful offered it as an option, and only one school failed to offer any music instruction. Most schools employed music specialists (94.9%), whereas some used classroom teachers (4.7%). The decision to employ a music specialist rested with the school board and/or superintendent (69.3%), the principal (24.9%), or a combination of both of these (5.4%). Four principals reported that a school committee/council was responsible for deciding whether to hire a music specialist. Contact hours for those that offered music instruction at the primary level (K–2nd grade) were as follows: less than 1/2 hour per week (4.2%), 1/2 hour to under 1 hour per week (54.7%), or 1 hour or more (39.3%). Contact hours at the intermediate level (3rd–5th grade) were under 1/2 hour (6.1%), 1/2 hour to under 1 hour (48.6%), or 1 hour or more (44.4%).

Abril, C. R., & Gault, B. M. (2006). The state of music in the elementary school: The principal's perspective. *Journal of Research in Music Education, 54*(1), 6–20.

EXAMPLE 2.2 USING DESCRIPTIVE STATISTICS TO REPORT RAW DATA

Research Question 1 investigated whether there would be differences in sight-singing scores when high school choristers with extensive training in movable solfège syllables (coupled with Curwen hand signs) used or did not use hand signs. Descriptive statistics revealed a mean score of 10.37 ($SD = 4.23$) for students' sight-singing scores with hand signs and a mean score of 10.84 ($SD = 3.96$) without hand signs.

McClung, A. C. (2008). Sight-singing scores of high school choristers with extensive training in movable solfège syllables and Curwen hand signs. *Journal of Research in Music Education, 56*(3), 255–266.

Means and standard deviations for secondary socialization influence on the decision to continue studying music

Source	Mean	SD
Applied studio teacher(s)	4.50	0.90
Parents	4.46	0.81
Taking lessons	4.42	0.98
Performing in ensembles	4.40	0.81
Attending college/ university performances	4.32	0.79
Attending classical music concerts	4.25	0.86
Informal music making outside of school	4.21	0.84
Ensemble director(s)	4.20	0.96
Other music students	4.17	0.91
Music education faculty	4.00	0.96
Other music faculty	3.95	0.91
Academic classes in music	3.93	0.96
Attending music festivals	3.87	0.88
Performing on recitals	3.87	0.97
Attending popular music concerts	3.83	0.86
Peers outside of music	3.75	0.93
Taking auditions	3.69	1.11
Attending music conferences	3.63	0.85
Teaching lessons	3.58	0.85
Field experience or internships	3.57	0.88
Faculty outside of music	3.46	0.81
Being a section leader	3.46	0.78
Academic classes outside of music	3.39	0.92

Note: 1 = very negative influence, 2 = somewhat negative influence, 3 = no influence, 4 = somewhat positive influence, 5 = very positive influence.

FIGURE 2.1

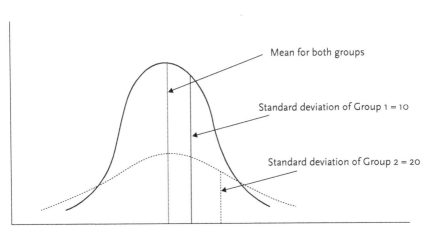

FIGURE 2.2

more clustered together, more so than the solid-line group. In this instance we see that the kurtosis (see the definition in Chapter 1) for the dashed-line group is greater than that of the solid-line group. What does this difference tell us? A larger standard deviation (like the one found with the solid-line group) is indicative of less agreement or similarity in responses (depending on the type of response). Although participants may have responded with or scored similar means between the two groups, there was less agreement in scores for the solid-line group. This is important data, which if omitted gives the reader too little information to make decisions about the findings of a study. If the data is omitted, the reader also will have less information to go upon to make sure that any subsequent inferential analyses are appropriate. As we will see in subsequent chapters, we need to make sure that we are comparing groups that share similar variances.

For example, a music education researcher may be interested in comparing two different groups of master's-degree students' intonation. Upon examining the means of the two groups, the researcher may find that the two groups have similar means (e.g., 7.8 and 7.6 on a 10-point scale). However, the standard deviation of the first group may be 1.2 and the standard deviation of the second group may be 3.1. These two variances may be different enough so that the researcher could not, with any level of confidence, tell if the groups are really the same or different outside the realm of chance.

Assumptions of Descriptive Statistics

Descriptive statistics do not really have assumptions of their own. The major issue with using descriptive statistics, beyond that of describing the data and research outcomes prior to more analytical or inferential procedures, is deciding which type of statistical procedure is most appropriate (i.e., parametric or nonparametric). If you hope to use parametric tests, the main assumptions of the descriptive data are that they are normally distributed and that they have relatively equal variances. We will learn later in this chapter different ways to test for normality. Examining equal variances is discussed throughout the book as it applies to each different statistical procedure.

Research Design and Descriptive Statistics

Although descriptive statistics are usually not associated with a specific research design, it is important to note the value of purely descriptive studies. Music education researchers should report descriptive statistics in any research project they conduct. However, sometime the descriptive statistics are the ultimate goal of a project. Without knowing what *is*, it can be difficult to know what questions we need to ask the next time around. Such descriptive studies are often in the form of a survey design or a one-shot case study.

Setting Up a Database for Descriptive Statistics

The examples given in this text focus on how to set up databases in SPSS (Statistical Package for the Social Sciences). However, database setup will be similar regardless of your preferred statistical software.

For these next sections, please use the provided descriptive-statistics dataset (Descriptive Example 1 ▶) to follow along and complete the analyses as you read. In this dataset, we can see four columns of data. We should set up columns for each individual variable so that we are able to have individual information for each collected variable for each participant on one single line (see Figure 2.3). These columns contain information that is both descriptive of the participants (gender, teaching genre, and teaching experience) and of their scores on a fictitious teaching evaluation. Gender is often considered a binary descriptor (i.e., male, female), although not always, and music education researchers should be respectful and offer more than binary choices to potential participants in regard to the socially constructed and reinforced construct of gender. In such cases, however, we create dummy variables in

	Gender	Genre	experience	Eval
1	0	1	5	5
2	0	2	12	5
3	0	1	21	5
4	1	2	4	4
5	1	3	1	5
6	1	3	1	4
7	1	1	2	5
8	1	2	3	1
9	0	3	9	3
10	1	3	15	1
11	1	3	6	2
12	1	3	27	1
13	1	3	29	1
14	1	2	31	1
15	0	2	1	5

FIGURE 2.3

which we score one gender, in this instance male, as 0 and the other, female, as 1. Participants could have responded one of three ways in regard to teaching genre (1 = Orchestra, 2 = Choir, and 3 = Band). Teaching experience has no established range and is entered just as participants offered the information. Finally, the teaching evaluation had a possible score range of 0 to 5.

Examining the Normality of Descriptive Statistics

Prior to reporting the descriptive statistics themselves, music education researchers may want to examine the normality of the data. Two of the most ubiquitous means of doing so is through a Kolmogorov-Smirnov test or a Shapiro-Wilk test, which are relatively easy to do through SPSS. What is the difference between the two? The Shapiro-Wilk test is better suited to analyzing data from a relatively small sample size. Generally speaking, if you have fewer than 50 participants, it is the better of the two for your analysis. Otherwise, the Kolmogorov-Smirnov test is quite adequate. To conduct both of these tests, follow these steps:

1. *Click* **Descriptive Statistics**.
2. *Click* **Explore**.
3. *Move the variable that you want tested into the dependent list (for our example, we will move* **Experience** *into the* **Dependent** *box).*
4. *Click* **Statistics**.
5. *Make sure to highlight the* **Descriptives** *box.*
6. *Click* **Continue**.
7. *Click* **Plots**.
8. *Make sure to select* **Normality plots with tests**.
9. *Click* **Continue**.
10. *Click* **OK**.

Once you have clicked **OK**, you will be provided with several boxes of data and plots. We are most interested in the first three boxes. The first box Figure 2.4) provides information regarding the *N* of the analysis.

The second box (Figure 2.5) provides us with the descriptive information about the data. We learn the mean, median, variance, standard deviation, range, and skewness and kurtosis. .

Case Processing Summary

	W					
	Valid		Missing		Total	
	N	Percent	N	Percent	N	Percent
experience	31	100.0%	0	0.0%	31	100.0%

FIGURE 2.4

Descriptives

			Statistic	Std. Error
experience	Mean		10.58	1.791
	95% Confidence Interval for Mean	Lower Bound	6.92	
		Upper Bound	14.24	
	5% Trimmed Mean		9.94	
	Median		6.00	
	Variance		99.385	
	Std. Deviation		9.969	
	Minimum		1	
	Maximum		32	
	Range		31	
	Interquartile Range		18	
	Skewness		.942	.421
	Kurtosis		−.441	.821

FIGURE 2.5

Test of Normality

	Kolmogorov-Smirnov[a]			Shapiro-Wilk		
	Statistic	df	Sig.	Statistic	df	sig.
experience	.225	31	.000	.841	31	.000

[a]Lilliefors Significance Correction

FIGURE 2.6

We could use this information to initially examine the normality of the data. Are the mean and median relatively similar? In our example, they are not. The mean is 10.58 and the mode is 6. This disparity does not bode well for a normal distribution. Another means of initially examining the normality of the data is by comparing the skewness statistic with the standard error of skewness. If you divide the skewness statistic by the standard error of skew and get a number that is less than −2, your data may be negatively skewed. If you obtain a number that is greater than 2, your data may be positively skewed. Although not perfect, this method can be used to strengthen an argument in either direction for the use of a particular type of inferential test to be conducted later. The same process holds true for kurtosis. The problem with this method, however, is that it is greatly impacted by the sample size. Therefore, researchers are usually better off relying on one of the tests that we see in Figure 2.6.

In Figure 2.6, we see the actual outcomes of the tests of normality. Because our sample size was only 31, it would be most appropriate to make our judgment on the basis of the Shapiro-Wilk test. Here we see the statistic for that test is .841 with 31 degrees of freedom and a significance level of $p < .001$. As the significance level for this test is smaller than .05, we know that we do not have a normally distributed dataset. If we were using this information as a primary variable in a subsequent inferential test, it would be most appropriate to use a nonparametric test.

Analyzing Nominal or Categorical Descriptive Statistics

For variables such as gender and genre, which are nominal or categorical (see definitions in Chapter 1), researchers will often report the frequencies of the information rather than means and standard deviations. In order to compute the frequencies in SPSS, follow these steps:

1. Click **Analyze**.
2. Click **Descriptive Statistics**.
3. Click **Frequencies**.
4. Move the categorical variables you want to examine (in this case, gender and genre) in to the Variables list by selecting them and clicking on the arrow button.
5. Click **OK**.

The readout you get should look something like that which we see in Figure 2.7.

In this figure, we see the three boxes that provide the information about these categorical variables. The first box provides the *N* of each examined variable, while the following two boxes provide the frequency counts of each

Statistics

		Gender	Genre
N	Valid	31	31
	Missing	0	0

Frequency Table

Gender

		Frequency	Percent	Valid Percent	Cumulative Percent
Valid	0	16	51.6	51.6	51.6
	1	15	48.4	48.4	100.0
	Total	31	100.0	100.0	

Genre

		Frequency	Percent	Valid Percent	Cumulative Percent
Valid	1	8	25.8	25.8	25.8
	2	11	35.5	35.5	61.3
	3	12	38.7	38.7	100.0
	Total	31	100.0	100.0	

FIGURE 2.7

variable. For example, we see that just over half (51.6%) of the participants in this dataset indicated that they were males ($n = 16$). What's the difference between percent and valid percent? If you were missing some data, the percent would be smaller, as it would tell you the percentage of the total people who answered, but the valid percent would take out responses from all those who did not answer all the items. The valid percent will always equal 100. The final box shows you that 8 people, or 25.8%, taught orchestra, while roughly equal numbers of participants taught choir or band (35.5% and 38.7%, respectively).

Before we move on to the other two columns of data, is it clear why we did not need to report the variance? These categorical or nominal datasets are not being compared to anything or even within themselves. Knowing the mean between males and females (in this case .48, rounded) tells us nothing beyond what the frequency and percentage tell us.

Analyzing Ratio or Scaled Descriptive Data

For the ratio data (see the definitions in Chapter 1) it is most common to report the mean and standard deviation. In order to compute the mean and standard deviation in SPSS, follow these steps:

1. Click **Analyze**.
2. Click **Descriptive Statistics**.
3. Click **Descriptives**.
4. Move the scaled or ratio variables you want to examine (in this example, **experience** and **eval**) in to the **Variables** list by selecting them and clicking on the arrow button.
5. Click on the **Options** button and make sure to select **Mean** and **Standard Deviation** are selected.
6. You can also elect to have other information provided if you wish (e.g., **Range**, **Skewness**, **Kurtosis**).
7. For ease of readability and reporting, click on **Descending Means** under the **Display Order** option.
8. Click **Continue**.
9. Click **OK**.

The readout you get should look something like Figure 2.8.

Descriptive Statistics

	N	Mean	Std. Deviation
experience	31	10.58	9.969
eval	31	3.32	1.514
Valid N (listwise)	31		

FIGURE 2.8

The first column of this table tells you that all 31 participants responded to both questions. The second column gives you the mean of each score, while the final column tells you the standard deviation of each item. From this information we know that the average amount of teaching experience for the participants was 10.6 years and that the average score on the teaching evaluation was 3.32 with a standard deviation of 1.51. If you would like to see how these results look in graph form with a normal curve, follow these steps:

1. *Click* **Graphs**.
2. *Click* **Legacy Dialogs**.
3. *Click* **Histogram**.
4. *Move the variable you would like graphed into the space labeled* **Variables** *(in this case,* **experience***)*.
5. *Click the box that reads* **Display normal curve**.
6. *Click* **OK**.

The graph you get would look like Figure 2.9.

Is the mean always enough information? No. As you can see in Figure 2.9, although the mean is 10.6, the large standard deviation is an artifact of the skewed data toward less experienced teachers. This is a prime example of when the mean would also be informed by the other indicators of central tendency: median and

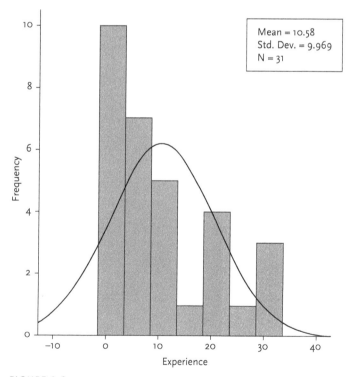

FIGURE 2.9

mode. In this example, the median is 6 and the mode is 1. These results help give a clearer picture of the experience level of this group of teachers.

Quick Hint

To compute the mode and median by using SPSS, follow these steps:

1. *Click* **Analyze**.
2. *Click* **Descriptives**.
3. *Click* **Frequencies**.
4. *Move the desired variables over the* **Variables** *list by using the arrow button.*
5. *Click on the* **Statistics** *button.*
6. *Highlight* **Median** *and* **Mode**.
7. Click **Continue**.
8. Click **OK**.

In Figure 2.10, you can see the graph with normal curve added for participants' evaluation score.

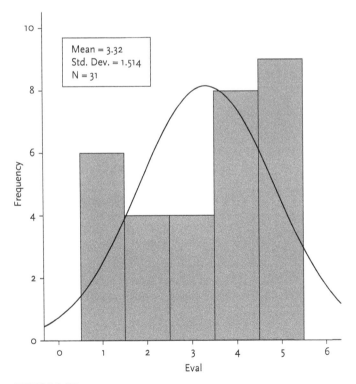

FIGURE 2.10

DESCRIPTIVE STATISTICS | 41

In Figure 2.10 we see that although there is a slight skew to the right, the normal curve looks more balanced than in Figure 2.9. If we examine the other indicators of central tendency, we will find that the mode is 5 and the median is 4. This result tells us that the most common score was 5 and the midpoint of all scores was 4. It is easy to see why the curve, although not perfectly aligned, looks more balanced.

Reporting Descriptive Statistics

What might a report of this information look like? See Example 2.3.

Writing Hints for Descriptive Statistics

From Example 2.3, we see a few writing hints that may help you state your findings as directly as possible.

- When you are discussing all of the participants, capitalize the N. When discussing a subgroup of your participants, lower-case the n.
- Per American Psychological Association guidelines, use first-person pronouns for clarity, especially when describing actions you took, while keeping all of your actions in past tense.
- Use the active voice in your sentences: Noun > Verb > Predicate.
- When you are comparing numbers lower than 10, as in this example, do not write them out.
- If there is a greater number of items to share (especially ones on similar scales), a table with descending means will be of benefit to the reader.

EXAMPLE 2.3 POSSIBLE DESCRIPTIVE DATA REPORT OF THE EXAMPLE DATASET

Participants ($N = 31$), of whom 51.6% were male, reported an average of 10.6 years of teaching experience ($SD = 9.97$). The majority of participants taught band (38.7%) or chorus (35.5%), while roughly a quarter taught orchestra (25.8%). At the end of the school year, each participant completed a teaching effectiveness evaluation with a possible score ranging from 1 to 5. The majority of participants ($n = 9$) scored a perfect 5, while six teachers scored a 1. Overall, the participants scored an average of 3.32 ($SD = 1.51$).

Final Thought on Descriptive Statistics

Descriptive statistics are often not given their due. Researchers should know the descriptive statistics for their datasets prior to moving on to any inferential analysis. Understanding the raw data is of paramount importance to understanding the data and making the most informed judgments about the data after all analyses are completed.

Descriptive Statistics Exercises

Given the second provided dataset, Descriptive Statistics Example 2 ⓘ, compute the most appropriate descriptive statistic for each column. In this example the columns include the following:

- Gender (0 = male, 1 = female)
- Teaching Genre (1 = orchestra, 2 = choir, 3 = band)
- Teaching Experience
- Teaching Evaluation
- Student Average (the average percentage the teacher awards his or her students)
- Student All-State (the number of students who made all-state during the year)

Once you have computed the most appropriate descriptive statistics, create a graph that includes the normal curve and describe the additional information to tell yourself about the statistics. Compute median and mode to see if this graph is helpful.

Finally, create a report of the statistics you found that offers the most pertinent information as succinctly as possible. Check your statistical work with the information found in the appendix to make sure you completed the analysis accurately.

II | Parametric Statistical Procedures

3 | Pearson Product-Moment Correlation

Use of Pearson Product-Moment Correlation Tests

Music education researchers can use correlation analysis in order to determine if a linear relationship exists between two variables and, if a relationship does exist, the strength and direction of the relationship. The variables in a correlation analysis should be continuous and not categorical; however, the two variables do not have to be measured on the same scale. For example, you can examine the relationships between teaching experience and sight-reading scores. Your teaching experience will most likely be on a scale of 0–50, while a sight-reading score may be on a scale of 0–5 or so. This disparity does not negatively impact the results of a correlation analysis.

Before we continue, there is one fundamental principle that we must keep in mind when thinking about and writing about correlational analyses. It is important to remember when you are conducting this type of analysis that a correlation between two variables *does not* show causation. It cannot be assumed that because two things are related that one thing caused the other. For example, a music education professor could ask graduate students in a large summer class to indicate the extent to which they have some skill in music, any skill at all on a 10-point scale. We could imagine that all those in the room would rate themselves relatively highly (as they should). Then the professor could ask the students how important it is to wear shoes in public. We could assume that this would be a relatively highly valued outcome as well. In this example, we would most likely find that there is a strong and direct correlation between wearing shoes and skill in music. Based on this outcome, we might conclude that wearing shoes leads to musical skill. This is an extreme and rather ridiculous example, but it illustrates the logic of the process of correlation. Relationships may exist and it is important to learn about relationships. Causation, however, cannot be discussed on the basis of any correlational analysis.

The most common parametric correlation analysis is a Pearson product-moment Correlation, which is denoted by the letter r. The relationship between two variables as described in a correlation can range from -1.00 to 1.00. A correlation of 0 means that the two variables are not related at all, while a -1.00 indicates a perfect indirect correlation (often referred to as a negative correlation) and a 1.00 indicates a perfect direct correlation (often referred to as a positive correlation). For the purposes of this book, we adopt the language of direct and indirect correlations rather than positive and negative correlations. The language that we use matters, and not all direct correlations are positive things. For example, a music education researcher who is interested in music-teacher well-being may be interested in a study that tracks alcohol consumption over the career of music teachers. The researcher may find a significant direct relationship between teaching experience and alcohol consumption. We already know that we cannot argue that teaching *causes* more drinking, but we can also probably agree that this relationship is not "positive." By adopting the more clinical verbiage, we help focus the reader and the analysis on the possible meanings of the analysis.

As shown in Figure 3.1, it is easy to examine correlations by using simple plot graphs. Two variables that are not related have no discernable linear relationship, while indirectly related variables have an inverse relationship (as one goes up, the other goes down) and directly correlated variables have a direct relationship (when one goes up, the other goes up; when one goes down, the other goes down).

These relationships, be they direct or indirect, also have a strength or magnitude. The closer to -1.00 or 1.00 that a relationship gets, the stronger the relationship between the two variables. As shown in Figure 3.2, as more data points near the line of best fit, the strength of the relationship increases. It is not about the angle of the line of best fit, rather about the number of data points that do not align.

Although no hard and fast rules exist for what is considered a strong relationship versus a weak relationship, researchers have established a de facto

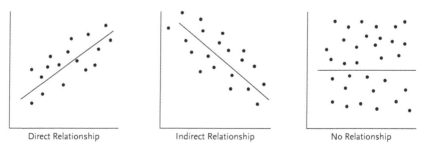

FIGURE 3.1

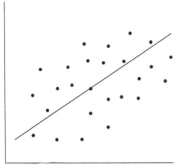 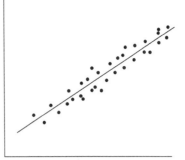

Moderate Correlation, approx. .45   Strong Correlation, approx. .85

FIGURE 3.2

EXAMPLE 3.1 REPORTING OF A PEARSON CORRELATION

Pearson product-moment correlations were calculated to determine how practice motivation and practice regulation subscale scores were related to instrument experience, frequency of practice, and amount of practice. A summary of those results appears in Table 3. Overall, the magnitude of the correlations was relatively weak. A small but significant negative correlation (–.16) emerged between practice motivation and band/orchestra instrument experience; students who had been playing their instruments for a greater length of time tended to express less motivation for practice. Practice motivation was positively correlated with both practice frequency and the amount of practice, but practice regulation correlations were stronger for both practice frequency ($r = .34$) and amount of practice ($r = .21$).

Berg, M. H., & Austin, J. R. (2006). Exploring music practice among sixth-grade band and orchestra students. *Psychology of Music, 34*(4), 535–558.

---

hierarchy of correlation strengths. Generally speaking, correlations ranging from 0 to .29 (or –.29) as weak, .30 to .69 (or –.30 to –.69) as moderate, and anything larger than .70 (or –.70) as a strong correlation. It is with these measurements that a music education researcher can discuss the practical significance of a finding once statistical significance has been found. See Examples 3.1 and 3.2.

In Example 3.1 we see an example of a researcher who overtly stated that the magnitude (aka strength) of the correlations is weak. It is a good thing for researchers to overtly discuss, as the discussion places the findings into more interpretable context, which is important. In Example 3.2, researchers have overtly discussed the practical significance of their findings on the basis of the magnitude of the relationships.

> **EXAMPLE 3.2 REPORTING PEARSON CORRELATION WITH PRACTICAL SIGNIFICANCE**
>
> We conducted a Pearson product-moment correlation to see if any individual difference variables (teaching level, teaching genre, age, teaching experience, degree level, certification, orchestra playing experience, number of string techniques courses, number of string pedagogy courses, or private lessons on a string instrument) related to the underlying perceived benefits or negative outcomes of string programs. Three factors were significantly related to possible benefit factors ($\alpha = .05$). Participants who had performed in an orchestra ($r = .120$) and who had taken more string techniques courses ($r = .143$) were more likely to believe that programs would benefit in logistical and support issues. Although these relationships were significant, the magnitude of the relationships are meager, indicating relatively little practical significance.
>
> ---
>
> Russell, J. A., & Hamann, D. L. (2011). The perceived impact of string programs on K–12 music programs. *String Research Journal, 2*, 49–66.

Assumptions of Pearson Correlation

The Pearson product-moment correlation has several assumptions that are similar to those of other parametric tests. They include the following:

1. The variables need to be continuous. Sometimes, researchers create "dummy" variables (i.e., group membership = 1, 2, 3, or 4) for categorical data to fit into a correlation analysis. Although this method can achieve a goal, it is more appropriate to use a categorical variable as an independent variable in an analysis of variance.
2. As with all parametric tests, the variables should be normally distributed.
3. Outliers (those data points that look nothing like the majority of data points) should be kept to a minimum. It is important to remove outliers (assuming that doing so does not negatively impact the other assumptions), as outliers can greatly influence the line of best fit, which can impact your findings.
4. The relationship between the two variables should be linear. That is, the relationship between two variables should remain constant rather than be nonlinear. This relationship can usually be most easily seen by using simple graphic representations of the data. See Figure 3.3.
5. There is homoscedasticity of the data. Homoscedasticity is similar to the assumption of homogeneity of variance. Simply put, it means that the variance between the two variables included in the analysis remains equal along the entire line of best fit. This concept is most easily tested using graphic scatterplots. See Figure 3.4.

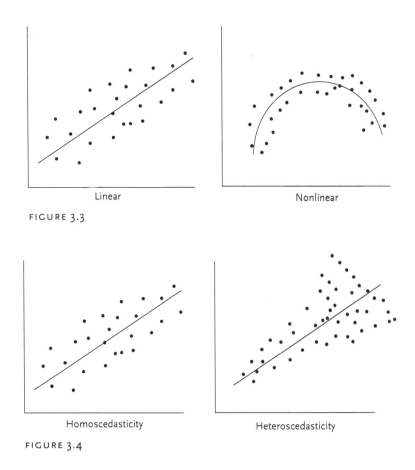

FIGURE 3.3

FIGURE 3.4

The Null Hypotheses for the Pearson Product-Moment Correlation

Although it is rare to overtly state the null hypothesis of the Pearson product-moment correlation, the null hypothesis would be that no relationship ($r = 0$) exists between the two variables. For example, a music education researcher may be interested in the relationship between students' achievement on a music sight-reading test and their grades in their English composition courses. The null hypothesis for this scenario would be that no relationship exists between the two variables. The alternative hypothesis would be that a relationship does exist between the two variables, although we do not know if the relationship is a direct one or an indirect one. Finally, a directional hypothesis would indicate that a relationship exists and that the relationship is either direct or indirect. For example, the researcher in our musical example might assume on the basis on previous research that a direct correlation exists between music sight-reading and the students' grades in English composition.

Research Design and Pearson Correlation

No research design really exists for the Pearson correlation. Most commonly, music education researchers use it to explore relationships between variables that are in addition to the main analysis of a study. Additionally, correlations are often used in descriptive studies and survey research studies, as in our musical example above in which the music education researcher is interested in examining any possible relationship between musical sight-reading and English composition scores. In this case, the researcher could use a Pearson correlation, if the data meets the assumptions, to see if indeed a relationship existed.

Setting Up a Database for a Pearson Correlation

The database for a Pearson correlation would have at least two columns of variables between which the researcher wants to examine possible relationships. Both of these columns would contain data from two continuous variables. Note that there has been no discussion of independent and dependent variables. In a bivariate (two variable) correlation analysis, the findings would be the exact same regardless of the order in which variables were analyzed (see Figure 3.5).

In Figure 3.6, we can see a database that has been created to examine the difference between high school music students' GPA and their scores on an individual performance final.

Conducting a Pearson Correlation in SPSS

For this section, please use the provided database, Pearson Example 1 ⏵, in order to conduct the tests while you are reading. This information will help

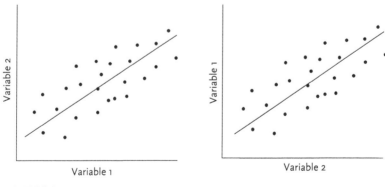

FIGURE 3.5

52 | Parametric Statistical Procedures

	GPA	PerformanceFinal
1	3.21	7
2	3.88	9
3	3.55	7
4	3.78	8
5	3.87	9
6	3.55	8
7	3.65	8
8	3.21	7
9	3.78	8
10	3.90	9
11	3.10	7
12	3.76	8
13	3.67	8
14	3.99	9
15	4.00	9

FIGURE 3.6

you obtain some guided experience in conducting and analyzing the data from a Pearson correlation.

Unfortunately, SPSS does not automatically create the scatterplots that make analyzing the data for meeting assumptions. So, the first thing that the researcher should do is create a scatterplot of the data in order to see if the data meets the assumptions. The assumption of normality can be assessed by using the methods discussed earlier in the chapter on descriptive statistics. The researcher needs the scatterplot to examine the homoscedasticity as well as look for an overabundance of outliers. In order to create a scatterplot, follow these steps:

1. Click **Graphs**.
2. Click **Legacy Dialogs**.
3. Click **Scatter/Dot**.
4. Select **Simple Scatter**.
5. Click **Define**.
6. Move one variable to the **Y Axis** slot and one to the **X Axis** slot. Remember that it does not matter which is which. In a correlation, there is no independent or dependent variable (see Figure 3.7).
7. Click **OK**.

Once you have clicked **OK**, you will get the graph shown in Figure 3.8.

From the scatterplot shown in Figure 3.8, you can see that there are no major outliers that would negatively impact the clarity of the analysis. Moreover, homoscedasticity seems to exist, given the relatively consistent shape of the line of best fit. On the basis of these findings, you should feel

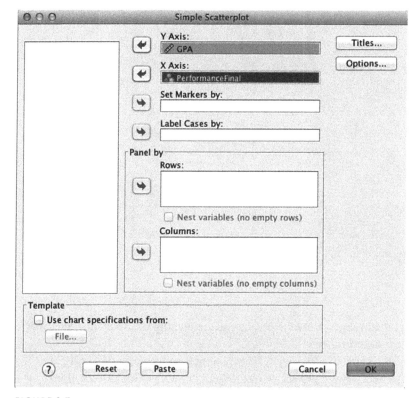

FIGURE 3.7

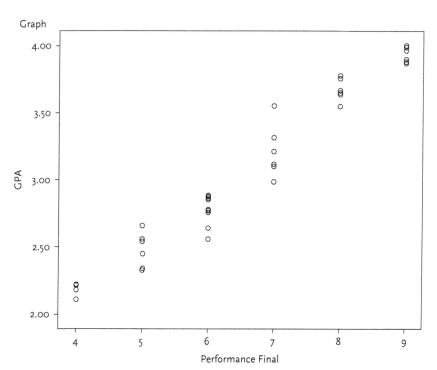

FIGURE 3.8

confident in continuing to conduct the Pearson product-moment correlation analysis.

In order to run the Pearson product-moment correlation in SPSS, follow these steps:

1. *Click* **Analyze**.
2. *Click* **Correlate**.
3. *Click* **Bivariate**.
4. *Move both variables over to the* **Variables** *list, making sure that you select* **Pearson** *under* **Correlation Coefficients**, *as shown in Figure 3.9.*
5. *Click* **Options**.
6. *Select* **Means** *and* **Standard Deviations**.
7. *Make sure that cases are excluded pairwise (this precaution keeps the largest number of participants possible in the analysis).*
8. *Click* **Continue**.
9. *Click* **OK**.

Once you have clicked **OK**, SPSS will provide you with two simple boxes. The first is just the descriptive statistics of the two variables (see Figure 3.10).

From this box, the researcher can see that 40 individuals participated in the study with a mean GPA of 3.09 with a standard deviation of .61. Similarly, participants scored an average of 6.65 on the performance final with a standard deviation of 1.59. This descriptive data is important and should be reported prior to reporting any inferential correlations that may or may not exist.

The second and final box that SPSS creates when a researcher conducts a Pearson product-moment correlation is the correlation matrix itself (see Figure 3.11).

From this box, you can see that the participants' GPA is directly correlated to their scores on the performance final. You can further see that all 40 participants were included in the analysis and that the significance of the relationship is $p < .001$ and that the strength or magnitude of the relationship is $r = .98$.

And yes, the matrix does repeat itself and compares variables to themselves. This phenomenon makes greater sense when you compare a larger number of variables, as it helps keep the matrix readable.

Is this enough information? Yes. From the analysis thus far, you can report all of the descriptive statistics, the discussion of meeting assumptions (if necessary), and the significance of the within-correlation analysis, as well as the strength of the relationship.

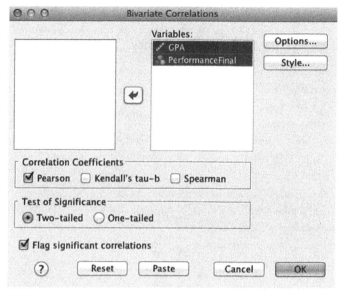

FIGURE 3.9

Descriptive Statistics

	Mean	Std. Deviation	N
GPA	3.0918	.61276	40
Performance Final	6.65	1.594	40

FIGURE 3.10

Correlations

		GPA	Performance Final
GPA	Pearson Correlation	1	.979**
	Sig.(2-tailed)		.000
	N	40	40
Performance Final	Pearson Correlation	.979**	1
	Sig.(2-tailed)	.000	
	N	40	40

**Correlation is significant at the 0.01 level (2-tailed).

FIGURE 3.11

Final Thought on Pearson Correlation

Do some researchers go too far on the basis of correlation? Yes. This point cannot be stressed enough: although the Pearson product-moment correlation is an apt tool to examine the relationship between two variables as well as the strength of that relationship, it cannot and does not imply causality. The researcher should not even hint at causality and might even warn readers to not infer any causality on the basis of a correlation analysis. Too many missteps in music education research have happened (and continue to happen) in no small part because of ignoring the simple fact of a relationship's existence does not mean that one caused the other. Post hoc, ergo propter hoc does not function in research. Just because something is related or follows in chronology, we cannot assume that it was caused by the other variable.

Reporting a Pearson Correlation

Once you have completed the Pearson analysis by using the provided dataset, the next step is to create the research report. Given the dataset in our running example, one possible write-up of the data may look like Example 3.3.

N.B. Although this is fabricated data, remember that although a relationship existed between their performance and GPA, it does not mean that one caused the other.

Writing Hints for a Pearson Product-Moment Correlation

From the research report in Example 3.3, we see some hints that might help you state your findings as clearly as possible while providing all of the necessary information.

- Make sure that you provide the following in your report:
  - N
  - r
  - p.
  - Discussion of practical and statistical significance (in the case of correlation, it is the strength of the correlation)
  - Means and standard deviations of all observations discussed

---

EXAMPLE 3.3 POSSIBLE REPORT OF A PEARSON CORRELATION

High school music students ($N = 40$) completed a final performance exam at the end of the school year. I collected this data as well as each participant's GPA. In order to examine if any relationship existed between the participants' GPA ($M = 3.09$, $SD = .61$) and their grades on the final performance evaluation ($M = 6.65$, $SD = 1.59$), I conducted a Pearson product-moment correlation and found a significant ($p < .001$) and strong ($r = .98$) relationship between the students' GPA and performance final, suggesting both statistical and practical significance regarding this finding.

- Per American Psychological Association guidelines, use first-person pronouns for clarity, especially when describing actions you took, while keeping all of your actions in past tense.
- Use the active voice.
- Avoid personification.
- Make sure all statistical indicators are in italics.
- Keep your actions as researcher in the past tense.

Pearson Product-Moment Correlation Exercises

Given the provided dataset, Pearson Correlation Example 2 ▶, create the scatterplots for each possible comparison and then compute the most logical analysis. In this example, a group of music educators in their first 10 years of teaching were assessed on their teaching effectiveness, asked to rate the joy they experience when teaching music, and given a music performance skill test. The columns are the following:

- Teaching Effectiveness on a scale of 1–10
- Teaching Experience (Years 1–10)
- Teaching Joy on a scale of 1–5
- Performance Skill on a scale of 1–5

Once you have computed the most appropriate statistical tests, write a report that includes all of the information required for a reader to assess the accuracy of your work, as well as the possible implications that may exist. Check your statistical work with the information found in the appendix to make sure you completed the analysis accurately.

4 | One-Sample T-Test

Use of the One-Sample T-Test

Although it is rather uncommon, at times researchers might want to know how data from a single group of participants in a study or population compares to that of either a known or a hypothesized mean. For example, a music education researcher may want to know if the participants from a survey of music teacher educators match the population from a previous study for comparison. That researcher could use a one-sample t-test to compare his data with the known data from a previous study to see if the means of the two sets of data are statistically significant or not.

As shown in Table 1.6 in Chapter 1, researchers using a one-sample t-test are trying to answer a basic research question: does a difference exist between the sampled group and a known mean?

Moreover, researchers can use a one-sample t-test to examine differences between the sampled group and a known or neutral/hypothesized mean. For example, a music education researcher may be interested in knowing if male students in her classes are more or less likely to play the trombone than one might hypothesize (i.e., one might hypothesize on the basis of experience or previous knowledge that the average all-state audition score of the trombone students would be 83). The researcher would be able to use the one-sample t-test to explore whether or not this is the case.

In Example 4.1 you can see that the researchers used a one-sample t-test to see if the means of their scores differed from a hypothesized neutral score. In this case, the neutral score was a score in the middle of their scale, or 3.5.

> **EXAMPLE 4.1 ONE-SAMPLE T-TEST USED TO DETERMINE DIFFERENCE BETWEEN SAMPLED MEAN AND NEUTRAL SCORE**
>
> The eight mean scores of the different scales of the IPSEE-T showed that teachers perceived the learning environment predominantly as a powerful learning environment. One-sample t-tests showed that on six of the eight scales the perception scores were significantly higher than the neutral score of 3.5 ($p < .01$ for all tests). Teachers perceived *fascinating contents*, emphasis on *productive learning, integration* in the learning contents, *interaction* during the learning process, *clarity of goals*, and *personalization* as significantly higher than neutral. On two scales the perception scores were significantly below 3.5 ($p < .01$ for both tests). Teachers perceived *differentiation* and *student autonomy* as significantly lower than neutral, although they were still scored higher than 3.0 (i.e., above the test choice "a bit disagree").
>
> ---
>
> Könings, K. D., Brand-Gruwel, S., & van Merriënboer, J. J. G. (2007). Teachers' perspective on innovations: Implications for educational design. *Teaching and Teacher Education, 23*, 985–997.

Assumptions of One-Sample T-Tests

The one-sample t-test has several assumptions that include the usual ones for most parametric statistical procedures. They are the following:

1. The data is continuous (i.e., interval or ratio, not nominal or ordinal). For example, scores on a music playing test, or responses to a scaled questionnaire item fit this criterion.
2. The data is not correlated and each observation is independent of any other observation.
3. The sample is normally distributed and unimodal (minimal skew or kurtosis; see Chapter 1).
4. The data should include no significant outliers.

Meeting the Assumptions of the One-Sample T-Test

Meeting the assumptions of the one-sample t-test is relatively easy. The first two assumptions listed above can be met by the design of the research being conducted. The remaining assumptions can be evaluated during data analysis. However, the one-sample t-test is relatively robust against the normal distribution assumption.

Research Design and the One-Sample T-Test

As the one-sample t-test is relatively uncommon, the designs in which it can be employed are often uncommon as well. This is not to say that these designs cannot be useful, but they are more rarely employed than most other parametric designs. Keep in mind that in the diagrams below, an O is an observation, X is a treatment of some sort, and an R stands for randomization.

(A) R X O
(B) X O
(C) O

As one can see, these three designs really represent the three levels of the same design. Design A is the experimental version and design B is the quasi-experimental version, while design C is a pre-experimental version of the design.

N.B. It may easily appear that these designs could be two-sample tests (e.g., independent-samples t-test). However, keep in mind that the one-sample t-test is used when you have actual data for one group, but only a mean or average for the comparison. If the research had the individual data for both groups, the most appropriate test would be the independent-samples t-test.

D.M. (*dignu m memoria*). Given the nature of these designs, a music education researcher will not be able to discern causality in any finding.

Let's take a look at what these designs might look like with a musical example.

(A) R $X_{\text{more time to practice prior to an audition}}$ $O_{\text{score on an audition}}$

In the example above, we see that the researcher was interested in whether or not the students who were given more time to prepare prior to an audition would score differently from the way typical students who completed the same audition did. A researcher could also collect data in two groups to compare by using an independent-samples t-test, but in this case, if we have only the known average score on the audition overall (e.g., 78), we can use that information in the one-sample t-test to see if the students who were randomly selected to be given more time to prepare achieve at the same level as those who were not.

(B) $X_{\text{music teachers undergoing writing/literacy professional development}}$ $O_{\text{teacher effectiveness ratings}}$

In this example (B), we see basically the same design as A, but lacking randomization. This may be a more beneficial design to researchers who work in schools where randomization can be challenging. In this example, a music supervisor in a large district is curious how the music teachers who attended a professional development program compare to the typical teacher on an end-of-year rating of teacher effectiveness. The music supervisor could use a one-sample t-test to examine if the means of the teachers who underwent the professional development were similar or different from the known score of all teachers in the district.

(C) O_{verbal sat score of music education program applicants}

In this example, the dean of a music school may be interested in comparing the trends in incoming students' SAT scores for the purposes of better understanding of her incoming students as well as for advocacy efforts to argue for increased scholarship allotments. If the dean knows the average verbal SAT score for all incoming first-year students at the institution, she would be able to compare the scores to see if the incoming music education students scored similarly or not on the test. Should these students have scored higher, she might be able to argue for greater financial support for these students based on academic merit.

The Null Hypothesis for the One-Sample T-Test

The null hypothesis for the one-sample t-test is that the mean of the observed group (μ_1) is equal to the mean of the known or hypothesized group ($\mu_{k/h}$). It could be stated as

$$H_0 = \mu_1 = \mu_{k/h}$$

where μ_1 = mean of the sample and $\mu_{k/h}$ = mean of the known or hypothesized group.

For our musical example above with regard to the verbal SAT scores of incoming music education students, the null hypothesis would be that the means of the observed group (incoming music education students) is equal to the mean of the known group (all incoming students from the university). It could be stated as

$$H_0 = \mu_{\text{verbal SAT score of all incoming music education students}} = \mu_{\text{verbal SAT score of all incoming music university students}}$$

The alternative hypothesis is that there would be a difference between the means of the incoming music education students and all students at the university. This alternative hypothesis could be written as

$$H_a = \mu_1 \neq \mu_{k/h}$$

In our musical example, the alternative hypothesis could be written as

$$H_0 = \mu_{\text{verbal SAT score of all incoming music education students}} \neq \mu_{\text{verbal SAT score of all incoming music university students}}$$

The dean of the school may have reason to believe that the average verbal SAT score of the incoming music education students is higher than that of the rest of the university's students. If that is the case, she may have a directional hypothesis, which can be stated as

$H_1 = \mu_1 > \mu_{k/h}$

$H_1 = \mu_{\text{verbal SAT score of all incoming music education students}} > \mu_{\text{verbal SAT score of all incoming music university students}}$

Should this directional hypothesis be confirmed by the one-sample t-test, the dean may be able to successfully argue for more academic merit scholarship for these students.

Setting Up a Database for the One-Sample T-Test

Setting up a database for a one-sample t-test is quite easy. It should just be a single column of data that acts as the dependent variable of the test. In Figure 4.1 we can see what a database should look like for this test. In this example, the researcher is interested in whether or not his students' audition scores are different from the average score from the previous year.

Conducting a One-Sample T-Test in SPSS

For this section of the chapter, please use the provided database (One-Sample t-Test Example 1 ▶) while working through the steps in SPSS. This step will help give the reader experience in conducting the test as a means of better understanding the process and interpretation of the data.

Conducting the one-sample t-test is one of the easier tests to complete in SPSS. In order to run the test, follow these steps in the program:

1. Click **Analyze**.
2. Click **Compare means**.
3. Click **One-Sample T Test**.
4. In the pop-up window (see Figure 4.2) select the dependent variable you want to analyze (in this case it is **Audition Score**).
5. Enter **Test Value**. This is the place where you enter the known or hypothesized value of the comparison. For this example, the researcher knows that the previous year's audition scores averaged 86. therefore, enter **86**.
6. Assuming you have data for each participant, you do not need to click on **Options**.
7. Click **OK**.

	AuditionScore
1	89
2	93
3	91
4	91
5	76
6	56
7	89
8	86
9	87
10	88
11	93
12	78
13	99
14	87
15	86

FIGURE 4.1

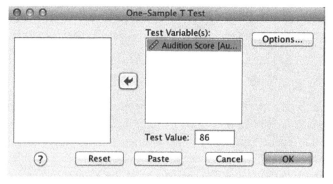

FIGURE 4.2

Once you have clicked **OK**, you should get only two boxes of information from SPSS. The first box (see Figure 4.3) is just the descriptive data. From Figure 4.3, we learn that there were 59 participants with a mean of 88.05 and a standard deviation of 6.36 and a standard error of .83.

The second box created by SPSS for this test holds the test outcome itself. From this box (see Figure 4.4), we can see the t-score, the degrees of freedom, the significance of the test, the mean difference, and the confidence interval of the difference. From this data, we can see that a significant difference did exist between the students from this year's audition to those who auditioned last year. Students this year scored two points higher overall, which was a significant difference ($p = .02$). This is really all the information you need to report the findings of a one-sample t-test.

One-Sample Statistics

	N	Mean	Std. Deviation	Std. Error Mean
Audition Score	59	88.05	6.361	.828

FIGURE 4.3

One-Sample Test

	Test Value = 86					
	t	df	Sig. (2-tailed)	Mean Difference	95% Confidence Interval of the Difference	
					Lower	Upper
Audition Score	2.476	58	.016	2.051	.39	3.71

FIGURE 4.4

Reporting a One-Sample T-Test

Once you have conducted the one-sample t-test in SPSS, the next step is to report the findings of the test in a manuscript. Given the analysis of the data examined in this chapter, the research report of this test may look like the one found in Example 4.2.

Writing Hints for a One-Sample T-Test

From the report found in Example 4.2, we can see some hints that might be helpful in making sure that your report of your findings is as clear as possible.

- Make sure that you provide the following in your report:
  - N
  - All of the descriptive statistics (mean and standard deviation)
  - All of the test data (t, df, p)
  - The mean difference
- Per American Psychological Association guidelines, use first-person pronouns for clarity, especially when describing actions you took, while keeping all of your actions in past tense.
- Use the active voice.
- Per APA guidelines, use first-person pronouns for clarity, especially when describing actions you took, while keeping all of your actions in past tense.
- Avoid personification.
- Make sure all statistical indicators are italicized.
- Make sure the reader knows that no causality can be inferred on the basis of the analysis.

EXAMPLE 4.2 POSSIBLE ONE-SAMPLE T-TEST RESEARCH REPORT BASED ON THE EXAMPLE 1 DATA

In order to examine whether or not students auditioning for all-state orchestra obtained similar or different scores than those from around the state, I conducted a one-sample t-test. I collected audition scores from each of my students who auditioned this year ($N = 59$) and knew the average score from the previous year (86) because of a public announcement from the festival chair. Initially, I examined the normality of the data and found no significant outliers and only one mode and deemed the data appropriate for analysis. Those who auditioned this year obtained an average audition score of 88.05 ($SD = 6.36$). On the basis of the analysis, I found a significant difference between this year's scores and last year's scores ($t = 2.48$, $df = 58$, $p = .02$). The mean difference was 2.05, which indicates that this difference may have practical significance as well. What remains unclear and requires additional research is the reason for this difference.

One-Sample T-Test Exercises

Given the second provided dataset, One-Way t-Test Example 2 ▶, complete the most logical analysis. In this example, the one column represents scores that students in a seventh-grade choir achieved on a music-notation reading quiz. The teacher is interested if her student achieved at the same level as students across the district. Obviously, the teacher does not have access to each student's scores, but she does know that the district average on this notational literacy test was 7.7.

Once you have conducted the test, write a report that includes any of the descriptive data necessary for understanding the test as well as to the test itself. Check your statistical work with the information found in the Appendix to make sure you completed the analysis accurately.

5 | Dependent-Samples T-Test

Use of the Dependent-Samples T-Test

Researchers use the dependent-samples t-test when they need to examine the differences in means between either the same group of individuals that has been tested twice (repeated measures design) or two groups of individuals that have been paired or matched. For this reason, the dependent-samples t-test is also often referred to as a paired-samples t-test. If the researcher is examining the pre-test and post-test scores of the same group, each participants would need to have data collected at each point. If, however, the researcher is using a paired sample, then instead of having each individual's score for both tests, he or she should use logical matching (i.e., similar backgrounds, experiences, age) to create pairs.

As seen in Table 1.6 in Chapter 1, researchers using a dependent-samples t-test are trying to answer a relatively simple question:, does a difference exist between two observations of the same phenomenon (paired samples) on one dependent variable?

For example, a music education researcher may be interested in tracking changes in music teachers' job satisfaction over the course of a year. To do so, the researcher might collect data on some scale in regard to job satisfaction at the beginning of a school year as well as at the end of the school year. As the researcher now has data from the same people at two different points, it would be appropriate to use a dependent-samples t-test to see if a significant change occurred from the start of the school year to the end of the school year.

When music education researchers use the dependent-samples t-test to examine differences between the means of the same group over a period of time, the dependent variable is derived from a research instrument that is used twice (either in the same form or in a parallel form that has been shown to be reliable with the first form of the test). The independent variable is usually some sort of treatment received between the pre-test and post-test. The same is true for a paired sample except that the researcher did not collect data twice from the same participants. In Example 5.1, researchers examine the pre-test

EXAMPLE 5.1 DEPENDENT SAMPLES T-TEST REPORT

Paired-samples t-test results revealed that all items had positive coefficients of correlation significant at the $p = .000$ level, confirming the appropriate use of this analysis procedure and an extremely strong relationship among participants in terms of direction of attitude changes. The t-tests revealed significant differences between pre-test and post-test scores on six items: two in General Attitudes and four in Music Attitudes. Students more strongly agreed on the post-test than on the pre-test with the statement "School is boring" (Pre M = 2.83, Post M = 2.57, $p = .028$). They less strongly agreed with the statements "I like being around other people" (Pre M = 1.51, Post M = 1.63, $p = .001$), "I like playing my instrument (singing)" (Pre M = 1.86, Post M = 2.09, $p = .001$), "Learning to play my instrument (sing) is important to me" (Pre M = 2.13, Post M = 2.41, $p = .000$), and "I will keep playing my instrument (singing) when I am an adult" (Pre M = 2.62, Post M = 2.77, $p = .041$). Finally, the students less strongly disagreed with the statement "Playing my instrument is boring" (Pre M = 3.93, Post M = 3.68, $p = .002$). Given the occasional differences between vocalists and instrumentalists identified in other analyses, paired-samples t-tests were run for the instrumental participants only. This analysis revealed that the significant differences identified for items 1, 5, 6, and 12 of Music Attitudes were due to changes in the attitudes of the PI participants. Table 3 includes all SI and PI means, standard deviations, and t-test significance levels for these four music-attitude items. The data indicates that the PI participants' valuing and enjoyment of playing their instrument dropped significantly (although the means still indicate relatively positive positions) while the SI participants' attitudes did not change.

Ester, D., & Turner, K. (2009). The impact of a school loaner-instrument program on the attitudes and achievement of low-income music students. *Contributions to Music Education, 36*(1), 53–71.

and post-test scores of an attitude instrument between students using school instruments (SI) and personal instruments (PI).

Notice in Example 5.1 that the researchers reported pre- and post-test means for each item as well as the significance of the difference (p). Also note that, in the report, the researchers discussed only the significant findings and reported the other necessary descriptive statistics in a table. The authors could have reported the standard deviation of each group (when in doubt, if you are reporting means, you should report standard deviations and maybe even skew and kurtosis) in text. This additional information helps the reader better understand the appropriateness of the data being used. In addition, the researchers could have reported the actual t-statistic, degrees of freedom (although this data could be derived from the information about participants), and effect size of the differences.

N.B. The degrees of freedom for a dependent-samples t-test is derived differently from the way the independent-samples t-test is. If you have 30 participants who each take a pre-test and a post-test, your degrees of freedom is 30 – 1 = 29. You do not count each observation as in the independent-samples t-test and, because there is only one group being compared, you subtract 1 rather than 2.

Assumptions of the Dependent-Samples T-Test

The dependent-samples t-test has the same five major assumptions as the independent samples t-test, which will be discussed in the next chapter:

1. The independent variable (or grouping variable) is bivariate. In the case of the dependent-samples t-test, that is to say, that the independent variable is usually pre-test and post-test scores from the same group of people or from paired groups of people.
2. The data collected for the dependent variable is continuous rather than categorical.
3. Each observation is independent of any other observation (most often accomplished through random sampling). Don't confuse this independence with having scores from the same person. In a dependent-samples t-test the scores of the pre-test and post-test are most likely related, but the score of one participant is not dependent upon the score of another.
4. The dependent variable is normally distributed (see Chapter 1).
5. The dependent variable has roughly the same variance in each group (variance is the standard deviation squared and often written as σ^2). Such consistency helps ensure that we are comparing apples to apples. This assumption is often referred to as homogeneity of variance. There is a slight difference here, though, for dependent-samples t-tests. The two "groups" are, most often, the same group, so what you are really looking for is homogeneity of variance between the two sets of scores rather than the two groups.

It is easier to argue for robustness against any violation of these assumptions if you have roughly equal numbers of participants or data points in each group and if there are more than 30 participants.

Meeting the Assumptions of a Dependent-Samples T-Test

You could argue that meeting all of the assumptions of a t-test is somewhat difficult. Researchers should strive, however, to conduct inferential statistical procedures only when it is appropriate. Otherwise, we are using poor information to inform our profession. In the case of the dependent-samples t-test, several of the assumptions are easy to accomplish. The bivariate nature of the independent variable is easy to meet and will be taken care of in the design phase of the study. The scores are usually a pre-test and post-test or "before and after." The dependent-variable data should be continuous, which is an easy assumption to meet as well and is considered long before data collection. Each observation should be independent from others. That is, the score of one participant should not be influenced by the score of any other participant, although the pre-test and post-test scores of individuals are most likely correlated. When we get to assumptions 4 and 5, normal distribution and homogeneity of variance, meeting assumptions can be a little more difficult. See Chapter 1 to review the methods for determining the normal distribution of a sample. Remember that you can use a histogram, or calculations of skewness and kurtosis, of your data to visually check the normal distribution and variance (which is the standard deviation squared) to compare variances.

Research Design and Dependent-Samples T-Test

The dependent-samples t-test can be used to compare the change over time or between the treatment of any group. So, the most ubiquitous research design that utilizes this type of inferential test is (O = an observation or data collection point, X is a treatment of any given type):

(A) O   X   O

Let's take a look at a musical example of this design in action.

(A) $O_{\text{music teacher job satisfaction}}$ $X_{\text{academic school year}}$ $O_{\text{music teacher job satisfaction}}$

In the design above, a music education researcher is interested in whether or not the rigors and joys of an academic school year have altered, for better or worse, the job satisfaction of music educators. In this design, the dependent-samples t-test is going to tell you if there was a change in mean between the pre-test and post-test for the one group (sample of music teachers) included in the design.

The Null Hypothesis for a Dependent-Samples T-Test

The null hypothesis in a dependent samples t-test is that the mean of the pretest will not be different than the mean of the posttest. It could be stated as

$H_0: \mu_{\text{pretest}} = \mu_{2 \text{posttest}}$

For our musical example above regarding the job satisfaction of music teachers, the null hypothesis would be that job satisfaction of teachers at the beginning of the school year would be the same as the job satisfaction of music teachers at the end of the school year. It could be stated as

$H_0: \mu_{\text{music teacher job satisfaction at start of year}} = \mu_{2 \text{ music teacher job satisfaction at end of year}}$

The alternative hypothesis would be that there would be a difference between the mean of the pre-test and post-test scores. It would be stated as

$H_a: \mu_{pretest} \neq \mu_{2\ posttest}$

In our musical example, the alternative hypothesis could be stated as

$H_a: \mu_{music\ teacher\ job\ satisfaction\ at\ start\ of\ year} \neq \mu_{2\ music\ teacher\ job\ satisfaction\ at\ end\ of\ year}$

Finally, the directional hypothesis would state that the mean of the pre-test and post-test would not be equal and that one would be higher or lower than the other. The most common directional hypothesis used with a dependent-samples t-test is that the post-test mean will be higher than the pre-test. This hypothesis would be stated as

$H_1 = \mu_{pretest} < \mu_{2\ posttest}$

In our musical example, should the researcher have some indication from previous research that music educators job satisfaction is lower at the end of the year, the directional hypothesis could be stated as

$H_a: \mu_{music\ teacher\ job\ satisfaction\ at\ start\ of\ year} > \mu_{2\ music\ teacher\ job\ satisfaction\ at\ end\ of\ year}$

If, however, the researcher had data to suggest that job satisfaction increased as the year progressed, the directional hypothesis could be stated as

$H_a: \mu_{music\ teacher\ job\ satisfaction\ at\ start\ of\ year} < \mu_{2\ music\ teacher\ job\ satisfaction\ at\ end\ of\ year}$

Setting Up a Database for a Dependent-Samples T-Test

The database for a dependent-samples t-test will have at least two columns for every dependent-samples t-test you will want to complete. In Figure 5.1, we can see a database in which three dependent-samples t-tests can be run. The first three columns are the pre-test scores of first-year music teachers' self-reported commitment to teaching, teaching enjoyment, and teaching job satisfaction. The following three columns include post-test scores for the same three topics taken at the end of the participants' first year of teaching.

Conducting a Dependent-Samples T-Test in SPSS

For this next section, please use the provided example database (Dependent Sample t-test Example 1 ▶) in order to go through the process while reading. This exercise may help you make greater meaning out of the reading and give you practice in running and interpreting the dependent-samples t-test.

	precommit	preenjoy	presat	postcommit	postjoy	postsat
1	9	8	9	9	8	3
2	8	7	8	8	7	4
3	9	8	9	7	8	5
4	8	9	8	9	9	3
5	7	7	7	8	8	4
6	7	8	9	7	8	5
7	7	7	8	9	7	3
8	9	9	7	8	8	4
9	7	7	9	7	8	4
10	8	8	8	9	7	4
11	7	7	10	8	8	4
12	6	6	9	7	9	3
13	7	7	8	9	7	5
14	6	9	9	8	8	4
15	8	8	8	7	7	3

FIGURE 5.1

In order to conduct the t-test in SPSS, follow these steps:

1. *Click* **Analyze**.
2. *Click* **Compare means**.
3. *Click* **Paired-Samples T Test**.
4. *Move the pre-test variable over to the first* **Paired Variables** *space (see Figure 5.2).*
5. *Move the post-test variable into the second* **Paired Variables** *space.*
6. *If you have multiple comparisons to run, you can enter more than one pair of variables at a time.*
7. *Click* **OK**.

N.B. If you run multiple comparisons such as this test, remember to consider the Type I error you might commit. In this instance, you as the researcher should conduct a Bonferroni correction (see Chapter 1). In this example, because there are three comparisons to make, you would alter alpha by dividing .05 by 3. You would then set the new significance level at .02. Another option that you could take is to automatically set alpha at the more conservative .01. However, in this case, that might increase a Type II error, as the adjusted .02 is more liberal. Striking a balance between the two types of error can be challenging. Therefore, it is helpful to always report what decisions are made with regard to error and why the you made the decisions.

Once you have clicked **OK**, you should see a readout with three boxes of information. The first box gives you the descriptive statistics that inform the following analyses. In this box (see Figure 5.3), you can see the mean, number of included participants (N), standard deviation, and standard error of the mean, that last of which estimates the error in predicting a population mean from a sample mean.

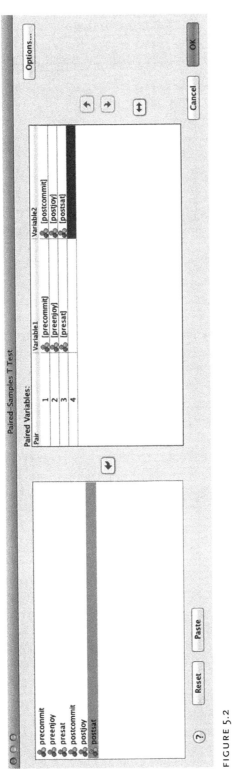

FIGURE 5.2

Paired Samples Statistics

		Mean	N	Std. Deviation	Std. Error Mean
Pair 1	pre commit	7.60	40	1.236	.195
	post commit	7.85	40	1.075	.170
Pair 2	pre enjoy	8.10	40	1.128	.178
	post joy	8.05	40	1.108	.175
Pair 3	pre sat	8.23	40	.920	.145
	post sat	4.10	40	.900	.142

FIGURE 5.3

From Figure 5.3, we can see that the means for two of the comparisons seem relatively stable from pre-test to post-test, that all 40 participants responded to each item, and that both the standard deviations and standard error remain relatively similar between the comparisons as well.

The second box that SPSS creates is an analysis of the correlation *within*, not between, the pairings (see Figure 5.4). This analysis tells us how correlated the pre-test and post-test scores are. There may be some confusion at this point. You might think that the correlation would be all that you need. If the scores are significantly correlated, there must be a significant difference. Not so. From the correlation information, the researcher can infer that one who did well on a pre-test, for example, did well on the post-test (if the correlation was positive). That inference does not assume, however, that the means of the two groups are significantly different.

From Figure 5.4, the researcher can see that only one correlation is significant: the pre-test of teacher commitment and the post-test of teacher commitment. As the correlation is direct, albeit moderate, the researcher knows that it is likely that individuals who reported high teacher commitment at the start of the year were likely to report high teacher commitment at the end of the year. The other two correlations, although not significant, indicate indirect relationships. This finding means that teacher enjoyment and teacher satisfaction went down (as evidence by the means in the first box, not intuitively).

From the third and final box, we can discover if any of the group means had a significant change from pre-test to post-test. See Figure 5.5.

What we are most interested in from this box (and what most needs reporting) is the right side of the box, including *t, df,* and the significance. The *t* statistic is reported for each item, as is the degrees of freedom (remember that the degrees of freedom for a dependent-samples t-test is the number of participants in the group minus one).

What does the formula for *t* look like for a dependent-samples t-test? It functions somewhat differently from the way the independent samples t-test

Paired Samples Correlations

		N	Correlation	Sig.
Pair 1	pre commit & post commit	40	.474	.002
Pair 2	pre enjoy & post joy	40	−.189	.243
Pair 3	pre set & post sat	40	−.214	.185

FIGURE 5.4

Paired Samples Test

		Paired Differences							
					95% Confidence Interval of the Difference				
		Mean	Std. Deviation	Std. Error Mean	Lower	Upper	t	df	Sig. (2-tailed)
Pair 1	pre commit- post commit	−.250	1.193	.189	−.632	.132	−1.325	39	.193
Pair 2	pre enjoy- post joy	.050	1.724	.273	−.501	.601	.183	39	.855
Pair 3	Pre set post sat	4.125	1.418	.224	3.672	4.578	18.403	39	.000

FIGURE 5.5

does, as we will see in the next chapter. The independent-samples t-test formula is

$$t = \frac{\overline{X}_{pre} - \overline{X}_{post}}{\sqrt{\dfrac{var_{pre}}{n_{pre}} + \dfrac{var_{post}}{n_{post}}}}$$

The formula for a dependent-samples t-test, however, is

$$t = \frac{d - \mu_d}{S_d \div \sqrt{n}}$$

You can see some specific differences as the dependent-samples t-test formula focuses more on the differences between the two observations of the same group rather than the means and is less concerned about the variance relationships (thus no Levene test is computed as the Levene test examines the difference in variance between two groups). In the formula for the dependent-samples t-test, the letter d is used to signify the difference. Therefore, the mean of the difference of each individual's score minus the mean of the difference divided by the standard deviation of the difference divided by the square root of the number of participants is how we compute d. The numerator is computed for you by SPSS and is listed in the final box of the information as the pooled mean. The standard deviation is also reported directly to the right of the pooled mean, and N equals 40. Therefore, for our example,

$$t = 4.125 \div 1.418 \div \text{ the square root of } 40\,(6.32).$$

Finally, in Figure 5.5, the researcher can see that only one pair experienced a significant difference: teacher satisfaction. Remember that normally the alpha would be set at .05. However, because we conducted three analyses, we lowered the alpha to .02 to mitigate a Type I error. So, there is only one t-test in this example that is significant: teacher satisfaction. As we look up to the first box with the descriptive statistics, we see that teacher satisfaction has significantly decreased from the beginning of the year to the end.

Is this enough information? Not quite yet. The researcher should also report the effect size of the dependent-samples t-test in order to discuss the practical significance of the findings. As will be the case for the independent-samples t-test in Chapter 6, the effect size for the dependent-samples t-test should be computed by using Cohen's d.

You can easily compute Cohen's d statistic by using the equation

$$d = \frac{\bar{X}_1 - \bar{X}_2}{s}$$

Another way to think of it is that

$$d = Pooled\ M \div SD$$

The pooled mean and the standard deviation are reported in the SPSS output under **Paired Differences.**

You could also compute the Cohen's d by using the reported t statistic and number of pairs (N) by using the formula

$$d = \frac{t}{\sqrt{N}}$$

What does the effect size tell the researcher? It tells us hypothesized difference between the two population means by using a measure of the magnitude of the difference expressed in standard deviation units. Cohen's d can be interpreted as how many standard deviations apart the two means happen to be.

There was only one significant difference found in the example above, that is, the only one for which we need to compute the effect size. Using the first equation, in which we divided the pooled mean by the standard deviation, we get 2.91 (4.125 ÷1.418). Using the second equation, you would get 2.95 (18.403 ÷ the square root of 39, or 6.24). As these are negligible difference, either equation is appropriate. What we see is that the magnitude of the difference is almost three standard deviations. Cohen believed that d around .2 to .3 was small, while those around .5 were medium and any above .8 were large;

> **EXAMPLE 5.2 POSSIBLE WRITE-UP OF A DEPENDENT-SAMPLES T-TEST**
>
> First-year music educators ($N = 40$) responded to three researcher-created items designed to elicit their commitment to teaching, how much they enjoy teaching, and their level of job satisfaction on a scale of 1 (no commitment, etc.) to 10 (great commitment, etc.). Participants completed these ipsative (that is, forced-choice) items at the beginning of their first year of teaching as well as at the end of their first year of teaching. In order to examine any differences that may have occurred after their initial year of teaching, the researcher conducted a series of dependent-samples t-tests. In order to mitigate the threat of Type I error caused by multiple comparisons, the researcher employed a Bonferroni adjustment and lowered alpha to .002 (.05 ÷ 3 comparisons). Participants reported no significant change in their commitment to teaching ($t = -1.33$, $df = 39$, $p = .19$) or their enjoyment of teaching ($t = .183$, $df = 39$, $p = .86$). Music teachers did, however, demonstrate a significant loss of job satisfaction ($t = 18.40$, $df = 39$, $p < .001$) from the beginning of the year ($M = 8.22$, $SD = .92$) to the end of the year ($M = 4.10$, $SD = .90$). In order to examine the effect size of this difference, the researcher computed Cohen's d and found an effect size of 2.91, suggesting practical as well as statistical significance.

we see that 2.91 is a rather large effect size, suggesting some practical significance to the finding.

Reporting a Dependent-Samples T-Test

Remember that our three categories of variables are music teacher commitment, enjoyment, and job satisfaction. Each of these categories consists of a report at the beginning of the year (pre-test) and at the end of the teachers' first year (post-test). Once you have completed your analysis, the final step is to create a research report of the information that could be read, understood, and evaluated by others. One possible report of the example data could be like the one found in Example 5.2.

Writing Hints for Dependent-Samples T-Test

From the research report in Example 5.2, we see some hints that might help you state your findings as clearly as possible.

- Per American Psychological Association guidelines, use first-person pronouns for clarity, especially when describing actions you took, while keeping all of your actions in past tense.

- Keep your reporting in the past tense.
- State your findings in the active voice.
- Remember to avoid personification. Results and studies do not "do" anything.
- Remember to place the statistical identifiers (e.g., *t, SD, M*) in italics. Do not forget to report the means and standard deviations for each inferential test. If there are more than three, as in Example 5.2, you might consider reporting means and standard deviations in a table.
- Report the *t, df,* and *p* for all of the analyses you conduct even if they are not significant.
- Avoid statements like "the difference between X and Y neared significance." A difference is either significant or not on the basis of your a priori established alpha.
- Remember to overtly report all of your manipulations of the data or expectations (i.e., Bonferroni).
- It is always a good idea to discuss how you feel the decisions you made will impact error.
- When in doubt, always report more information. It is always better to report more than your readers need rather than less than they need.

Final Thought of Dependent-Samples T-Test

The dependent-samples t-test is becoming less ubiquitous in educational research, as we often have a hard time creating situations that allow for multiple groups experiencing different treatments that we can test by using two data points. However, researchers can use the dependent-samples t-test in well-designed research as well as in pre- or quasi-experimental designs when they want to objectively examine participant growth or understanding over time.

Dependent-Samples T-Test Exercises

Given the second provided dataset, Dependent Samples t-test Example 2 ⊙, compute the most logical analysis. In this example, the columns include two pairs of tests:

- Pretest and Posttest for a sight-reading test (SR) with a scale of 0 to 10.
- Pretest and Posttest for music aptitude test (MA) with a scale of 0 to 50.

Conduct an analysis of the two paired samples. Once you have computed the most appropriate statistical test, write a report that includes the descriptive statistics necessary to understand the dependent-samples t-test as well as the test itself and, finally, effect size. Check your statistical work with the information found in the appendix to make sure you completed the analysis accurately.

6 | Independent-Samples T-Test

Use of the Independent-Samples T-Test

Music education researchers use the independent-samples t-test when they want to compare the means of two (and only two) separate groups or levels of an independent variable. For example, a music education researcher may be interested in the differences between sight-singing scores of his select choir and his general choir. Our music education researcher could use an independent-samples t-test to find out if a significant difference existed.

As seen in Table 1.6 in Chapter 1, when music education researchers use an independent samples t-test, they are trying to answer a specific research question: does a difference exist between two groups' means on one dependent variable?

When researchers use a t-test to examine the differences between the means of two separate groups on a given research instrument, the dependent variable is derived from the research instrument and the independent variable is the group. For our choral example, the dependent variable is the scores on the sight-singing test, while the independent variable is the grouping (i.e., select choir or general choir).

In Example 6.1, we can see how some researchers used an independent-samples t-test to examine differences between groups.

Notice how in Example 6.1 the authors provide the means for each group (FC and FI) as well as the final significance (p). This information helps the reader situate the inferential statistic with the descriptive means. The authors could also have provided the standard deviations of the groups as well as the t-statistic itself. If the authors have stated the number of participants in each group, it is helpful, but not necessary, to report the degrees of freedom because with a t-test (two groups) there will always be $N - 2$ degrees of freedom. Remember, there are two independent observations in the independent-samples t-test, so instead of the $N - 1$ you might expect, we calculate the degrees of freedom as $N - 2$.

> **EXAMPLE 6.1 USE OF A T-TEST TO INDICATE DIFFERENCES BETWEEN GROUPS**
>
> The data from pre-tests and post-tests of both the FC and the FI group were analyzed to further determine the equivalency of the two groups. An independent-samples t-test was calculated between the pre-test scores of the two groups, as well as between the post-test scores. No significant differences were found between the FC and FI groups' pretest knowledge (FC $M = 63.65$, FI $M = 60.38$, $p = .25$), comfort (FC $M = 49.27$, FI M $= 41.96$, $p = .06$), and frequency of use (FC M $= 38.49$, FI $M = 38.08$, $p = .52$) scores. Likewise, examination of the post-test scores of the FC and FI groups revealed no significant differences in the participants' knowledge (FC $M = 81.43$, FI M $= 81.22$, $p = .93$), comfort (FC $M = 81.68$, FI $M = 77.21$, $p = .06$), and frequency of use (FC $M = 69.19$, FI $M = 67.17$, $p = .08$).
>
> ---
>
> Bauer, W. I., Reese, S., & McAllister, P. A. (2003). Transforming music teaching via technology: The role of professional development. *Journal of Research in Music Education, 51*(4), 289–301.

> **EXAMPLE 6.2 REPORTING OF INDEPENDENT SAMPLES T-TEST**
>
> An independent t-test revealed that students chosen for the band ($M = 23.14$, $SD = 30.86$) traveled a significantly shorter distance than those who were not chosen ($M = 36.32$, $SD = 36.65$) ($t = 2.31$, $df = 230$, $p < .02$). On the other hand, the difference in enrollment per grade between the successful group ($M = 228.41$, $SD = 177.22$) and the unsuccessful group ($M = 177.23$, $SD = 168.40$) was not significant ($t = 1.87$, $df = 230$, $p < .06$).
>
> ---
>
> Lien, J. L., & Humphreys, J. T. (2001). Relationships among selected variables in the South Dakota all-state band auditions. *Journal of Research in Music Education, 49*(2), 146–155.

In Examples 6.2 and 6.3, authors of a study provide the reader with the actual t-statistic along with descriptive statistics, degrees of freedom, and overall significance.

Assumptions of the Independent-Samples T-Test

The independent samples t-test has five major assumptions:

1. The independent variable (or grouping variable) is bivariate. That is, the independent variable is membership in one or another group or category.

> **EXAMPLE 6.3 REPORTING OF AN INDEPENDENT-SAMPLES T-TEST**
>
> The subjects spent an average of 81.2 minutes working with the instructional software, with a standard deviation of 12.9 minutes. The DM group spent an average of 86 minutes working with the instructional software, while the EM group spent an average of 75.5 minutes. The DM group spent more time than the EM group on each of the four lessons. Independent-samples t-tests revealed that the difference between the groups was statistically significant for total time spent in working with the software, $t(25) = 2.25$, $p = .03$. The DM group spent 25.2 minutes on Lesson 1 (ornamental), 20.7 minutes on Lesson 2 (figural), 21.8 minutes on Lesson 3 (tempo), and 18.3 minutes on Lesson 4 (modal). The EM group spent 22 minutes on Lesson 1, 16.7 minutes on Lesson 2, 19.5 minutes on Lesson 3, and 17.3 minutes on Lesson 4.
>
> ---
>
> Hopkins, M. T. (2002). The effects of computer-based expository and discovery methods of instruction on aural recognition of music concepts. *Journal of Research in Music Education, 50*(2), 131.

For example, one student could not be counted in both the select choir *and* the general choir in our example.

2. The data collected for the dependent variable is continuous rather than categorical.
3. Each observation is independent of any other observation (most often accomplished through random sampling).
4. The dependent variable is normally distributed.
5. The dependent variable has roughly the same variance in each group (variance is σ^2). Such consistency helps ensure that we are comparing apples to apples. This assumption is often referred to as homogeneity of variance.

Although it is not an assumption, it is easier to mitigate violations of these assumptions if you have equal numbers of participants or data points in each group and if there are more than 30 data points in each group.

Meeting the Assumptions of an Independent-Samples T-Test

As stated in Chapter 5 in regard to the dependent-samples t-test, you could argue that meeting all of the assumptions of a t-test is impossible. Nonetheless, researchers should strive to conduct inferential statistical procedures only when it is appropriate. Otherwise, we are using poor information to inform our profession and instruction. In the case of the t-test, several of the assumptions are easy to meet on the basis of the research design itself. The bivariate nature of the independent variable is easy to meet. If the independent variable is not

> **EXAMPLE 6.4 REPORTING THE USE OF LEVENE'S TEST FOR EQUALITY OF VARIANCE**
>
> Given the high correlations between the PMMA subtests and composite, only the composite scores were compared for school settings and grade levels. The assumption of homogeneity of variance was met for the composite test among the urban, suburban, and rural school settings (Levene's $F = 1.74, p > .05$).
>
> ---
>
> Stamou, L. (2010). Standardization of the Gordon primary measures of music audiation in Greece. *Journal of Research in Music Education, 58*(1), 75–89.
>
> Two of the differences were dismissed due to failure of Levene's test for homogeneity of variance.
>
> ---
>
> Ester, D. (2009). The impact of a school loaner-instrument program on the attitudes and achievement of low-income music students. *Contributions to Music Education, 36*(1), 53–71.

bivariate, you cannot use a t-test. The dependent-variable data should be continuous, which is an easy assumption to meet as well. Each observation should be independent from others. That is, the score of one participant should not be influenced by the score of any other participant. When we get to assumptions 4 and 5, normal distribution and homogeneity of variance, meeting assumptions is a little more difficult. See Chapter 2 on descriptive statistics to see how to examine the normality of your data. Finally, we come to the assumption of homogeneity of variance. For a t-test, this assumption is usually examined by using the Levene test for equality of variance, which is another way of saying homogeneity of variance). See Example 6.4 for some reporting examples.

In the first part of Example 6.4, the author reported the F statistic as well as significance level of the Levene test. Note that the author reported that the significance (p) was greater than .05. Remember that in this instance it is desirable for the two groups to be similar; therefore, a significance finding (one that is less than .05) means that the assumption of homogeneity of variance was violated.

Please note that in the second part of Example 6.4, the author made the ethical decision to eliminate variables that did not meet the assumption of homogeneity of variance.

Research Design and Independent-Samples T-Test

The independent-samples t-test can be used to compare any two groups that are independent of each other. So, the possible research designs that utilize this type of inferential test include

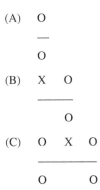

Let's take a look at a few musical examples of how these designs may be used by music education researchers.

(A) $O_{\text{select choir sight-singing scores}}$

   _____

   $O_{\text{general choir sight-singing scores}}$

In this example, as discussed above, the music education researcher is interested in whether or not a significant difference exists between the two different choirs on the basis of a single observation of sight-singing skill. There is no treatment in this design, although you could argue that experiencing the different choral settings is a treatment in itself.

(B) $X_{\text{sequencing technology instruction}}$ $O_{\text{composition project outcome}}$

   _____

   $O_{\text{composition project outcome}}$

In this example, the researcher was interested in whether or not instruction on using a new piece of music sequencing technology improved high school theory students' composition project scores. One group received the instruction, while the other did not. This test would be a way to see if the instruction itself was valuable, although it will be hard to make too many conclusions, as we did not know where the students were in regard to composition skill prior to the instruction.

(C) $O_{\text{audition score pre}}$ $X_{\text{anxiety mitigation training}}$ $O_{\text{audition score post}}$

   _____

   $O_{\text{audition score pre}}$ $O_{\text{audition score post}}$

This is the classic pre-test/post-test control-group design. In this example, the researcher was interested in the impact of anxiety mitigation training on the audition scores of the participants; the directional hypothesis was that those who learned to better calm their nerves would show greater gains in the high-pressure setting of audition scores. The independent-samples t-test is used only in this example (pre-test/-post-test design) if you are analyzing the pre-test of both groups, the post-test of both groups, or (and more accurately) the mean gain score of both groups. In this example, the most parsimonious analysis would be with the mean gain score for each group acting as the dependent variable, as the independent-samples t-test has only one dependent variable. Another possible analysis for this type of design is using the scores of the pre-test as a covariate. See Chapter 7 and 11 for more information about the analysis of covariance.

The Null Hypothesis for an Independent-Samples T-Test

The null hypothesis for an independent-samples t-test is that no difference will exist between the two groups. The most common way to state the null hypothesis tested by the independent-samples t-test is

$H_0: \mu_1 = \mu_2$

For our musical example above about sight-singing scores, the null hypothesis would be that no difference exists between the select choir and the general choir in regard to their sight-singing skill. This hypothesis could be stated as

$H_0: \mu_{\text{select choir sight-singing scores}} = \mu_{\text{general choir sight-singing scores}}$

The alternative hypothesis for the independent-samples t-test is that a difference does exist between the two groups. This hypothesis could be stated as

$H_a: \mu_1 \neq \mu_2$

For our musical example, the alternative hypothesis would be that a difference does exist in sight-singing scores between the select choir and the general choir participants. It could be stated as

$H_a: \mu_{\text{select choir sight-singing scores}} \neq \mu_{\text{general choir sight-singing scores}}$

Finally, should the music education researcher have evidence from previous research, the researcher may use a directional hypothesis. For the independent-samples t-test, the directional hypothesis would be

$H_1: \mu_1 < \mu_2$

In our musical example, the directional hypothesis may be that students in the select choir will have better scores on their sight-singing assessment than the students in the general choir. If so, this directional hypothesis could be stated as

H_1: $\mu_{\text{select choir sight-singing scores}} > \mu_{\text{general choir sight-singing scores}}$

Setting Up a Database for an Independent-Samples T-Test

The database for an independent-samples t-test needs only two columns. One column contains the dependent variable (e.g., scores on a sight-singing test). The other column contains the grouping variable (e.g., 1 = select choir and 2 = general choir). See Figure 6.1.

Conducting the Independent-Samples T-Test in SPSS

For this section, please use the provided dataset (Independent Sample T-test Example 1 ▶ in order to complete the analysis as you read. This exercise will give you some experience conducting and analyzing the data while you read through the chapter.

In order to conduct the t-test in SPSS, follow these steps:

1. Click **Analyze**.
2. Click **Compare means**.
3. Click **Independent-Samples T Test**.
4. Move the dependent variable over to the **Test Variable** *space*
5. Move the independent variable into the **Grouping Variable** *space*.
6. Click the **Define Groups** button.
7. Indicate that the group one will be represented by the value of 1 and the group two will be represented by the value of 2. All you are doing here is telling SPSS how you labeled your two groups. See Figure 6.2.
8. Click **Continue**.
9. Click **OK**.

	SightSingTest	Group
1	1	1
2	2	2
3	2	1
4	1	2
5	3	1
6	9	2
7	4	1
8	2	2
9	3	1
10	5	2
11	2	1
12	3	2
13	2	1
14	3	2
15	1	1

FIGURE 6.1

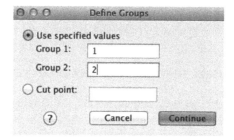

FIGURE 6.2

Group Statistics

	Group	N	Mean	Std. Deviation	Std. Error Mean
Sight Sing Test	1	25	3.40	1.384	.277
	2	25	3.92	1.801	.360

FIGURE 6.3

Once you have clicked **OK**, you will get an output with two boxes. The first box (see Figure 6.3) is important because it gives you the descriptive statistics, as discussed in Chapter 2.

In this box, we can see that we have equal numbers in each group ($n = 25$) as well as the mean for each group ($M_1 = 3.40$; $M_2 = 3.92$). We also can see the standard deviations of each group ($SD_1 = 1.38$; $SD_2 = 1.80$). The final column in this box is the standard error of the mean.

The second box that SPSS creates when you run the t-test has the inferential statistical information. See Figure 6.4.

From this figure, we can see in the first two columns under the Levene's test for equality of variance that the F statistic for this analysis is .686 and the significance (sig, p, α) is .411. Remember that we hope to *not* have a significant finding here, as we want both groups to be the same. We see that .411 is *not* significant (that is, it is above .05). Therefore, we can assume equal variances; thus we will use the upper row of information.

Had the Levene test shown that the groups had unequal variance, we would have used the lower row. Notice that the lower row has a smaller degree of freedom: SPSS has ejected outliers that created unequal variance. Ejecting outliers reduces power in a researcher's analysis, as this step removes participants and increases the threat of a Type I error. It is up to you as the researcher to decide whether or not you should continue with the analysis. The ethical thing to do, however, is to adequately report all of your decisions as well as the statistical reasoning behind them.

The next column gives us the actual t-statistic. The formula for the t-test is actually relatively simple:

Independent Sample Test

		Levene's Test for Equality of Variances		t-test for Equality of Means					95% Confidence Interval of the Difference	
		F	Sig.	t	df	Sig. (2–ailed)	Mean Difference	Std. Error Difference	Lower	Upper
Sight Sing Test	Equal variances assumed	.686	.411	−1.145	48	.258	−.520	.454	−1.433	.393
	Equal variances not assumed			−1.145	45.024	.258	−.520	.454	−1.435	.395

FIGURE 6.4

$$t = \frac{\bar{X}_1 - \bar{X}_2}{\sqrt{\dfrac{var_1}{n_1} + \dfrac{var_2}{n_2}}}$$

In this formula, the numerator is the mean of one group (in this case a group labeled 1) minus the mean of the second group (labeled 2).

The denominator is the variance of the first group (remember that variance is the square of the standard deviation) divided by the n in that group added to the variance of the second group divided by the number in that group. Once you have these two numbers, you compute the square root and then divide the numerator by the denominator, giving you t. As you can see from the logic in this formula, t is the difference between means while you take into account the variance (error) for each group. On the basis of this formula, you can see that the t-statistic could be either positive or negative, depending on which direction the change is made (this polarity has great implication when you are conducting a one-tailed test, but that point is for another day).

We can begin to examine the significance of the difference once you find t and know the degrees of freedom as offered in the next column (in the case of the independent-samples t-test, this is the number of total observations minus 2 ($N - 2$), as there were two observations. In Figure 6.4 we see that there are 48 degrees of freedom, because there were 25 participants in each group ($25 + 25 = 50 - 2 = 48$ df).

The final imperative piece of information needed is the significance column. This is the information that compares the t-statistic and degrees of freedom. Prior to programs such as SPSS, you would have to compute t and df and then check a t-table to find the p value for your data. See Figure 6.5 for a t-table.

With our example t-value of -1.15 we can check the t-table to see what our significance may be. We have 48 degrees of freedom, so we head to the bottom row. Now, don't be confused by the negative t-statistic. We are at this stage worried only about the absolute value of t. We see that 1.15 falls between the last two columns, which indicate a two-tailed significance of somewhere between .20 and .50. How does that match the SPSS output?

Given the data in Figure 6.4, we see that we have a significance level of .258 (the t-table worked!). As this number is greater than .05, we should accept the null hypothesis that the means of the two groups are indeed the same.

Is this enough information? Not quite. Although many researchers stop at reporting the statistical significance of the t-test, more information would be most helpful to determine the practical significance of the finding. For the t-test, one of the most common means of reporting effect size is using Cohen's d. This is a straightforward formula that will take you only a few moments to

df/p	0.40	0.25	0.10	0.05	0.025	0.01	0.005	0.0005
1	0.324920	1.000000	3.077684	6.313752	12.70620	31.82052	63.65674	636.6192
2	0.288675	0.816497	1.885618	2.909986	4.30265	6.96456	9.92484	31.5991
3	0.276671	0.764892	1.637744	2.353363	3.18245	4.54070	5.84091	12.9240
4	0.270722	0.740697	1.533206	2.131847	2.77645	3.74695	4.60409	8.6103
5	0.267181	0.726687	1.475884	2.015048	2.57058	3.36493	4.03214	6.8688
6	0.264835	0.717558	1.439756	2.015048	2.57058	3.36493	4.03214	6.8688
7	0.263167	0.711142	1.414924	1.894579	2.36462	2.99795	3.49948	5.4079
8	0.261921	0.706387	1.396815	1.859548	2.30600	2.89646	3.35539	5.0413
9	0.260955	0.702722	1.383029	1.833113	2.26216	2.82144	3.24984	4.7809
10	0.260185	0.699812	1.372184	1.812461	2.22814	2.76377	3.16927	4.5869
11	0.259556	0.697445	1.363430	1.795885	2.20099	2.71808	3.10581	4.4370
12	0.259033	0.695483	1.356217	1.782288	2.17881	2.68100	3.05454	43178
13	0.258591	0.693829	1.350171	1.770933	2.16037	2.65031	3.01228	4.2208
14	0.258213	0.92417	1.345030	1.761310	2.14479	2.62449	2.97684	4.1405
15	0.257885	0.691197	1.340606	1.753050	2.13145	2.60248	2.94671	4.0728
16	0.257599	0.690132	1.336757	1.745884	2.11991	2.58349	2.92078	4.0150
17	0.257347	0.689195	1.333379	1.739607	2.10982	2.56693	2.89823	3.9651
18	0.257123	0.688364	1.330391	1.734064	2.10092	2.55238	2.87844	3.9216
19	0.256923	0.687621	1.327728	1.729133	2.09302	2.53948	2.86093	3.8834
20	0.256743	0.686954	1.325341	1.724718	2.08596	2.52798	2.84534	3.8495
21	0.256580	0.686352	1.323188	1.720743	2.07961	2.51765	2.83136	3.8193
22	0.256432	0.685805	1.321237	1.717144	2.07387	2.50832	2.81876	3.7921
23	0.256297	0.685306	1.319460	1.713872	2.06866	2.49987	2.80734	3.7676
24	0.256173	0.684850	1.317836	1.710882	2.06390	2.49216	2.79694	3.7454
25	0.256060	0.684430	1.316345	1.708141	2.05954	2.48511	2.78744	3.7251
26	0.255955	0.684043	1.314972	1.705618	2.05553	2.47863	2.77871	3.7066
27	0.255858	0.683685	1.313703	1.703288	2.05183	2.47266	2.77068	3.6896
28	0.255768	0.683353	1.312527	1.701131	2.04841	2.46714	2.76326	3.6739
29	0.255684	0.683044	1.311434	1.699127	2.04523	2.46202	2.75639	3.6594
30	0.255605	0.682756	1.310415	1.697261	2.04227	2.45726	2.75000	3.6460
z	0.253347	0.674490	1.281552	1.644854	1.95996	2.32635	2.57583	3.2905
CI	—	—	80%	90%	95%	98%	99%	99.9%

FIGURE 6.5

compute, but will give your analysis much greater depth and meaning. As we saw in Chapter 5, the formula for Cohen's d is

$$d = \frac{\overline{X}_1 - \overline{X}_2}{s}$$

Therefore, d is the mean of the second group subtracted from the first group, then that outcome divided by the pooled (both groups) standard deviation.

> **EXAMPLE 6.5 REPORTING OF COHEN'S *d***
>
> The researches performed tests, resulting in significant differences ($p < .01$) between current and ideal states of music education for all variables under investigation. The magnitude of the effect was calculated for each variable by using a Cohen's *d* value. The variables that had a large effect size were "understanding music in relation to other subjects" ($d = 1.10$); "creating and composing music" ($d = 1.04$); "analyzing, evaluating and describing music verbally and in writing" ($d = .97$); and "understanding music in relation to culture and history" ($d = .86$). The following variables had medium effect sizes: "listen to music attentively," "read and write musical notation," and "perform music."
>
> ---
>
> Abril, C. R., & Gault, B. M. (2006). The state of music in the elementary school: The principal's perspective. *Journal of Research in Music Education, 54*(1), 6–20.

What is a small, medium, or large effect size? Cohen believed that *d* around .2 to .3 was small, while those around .5 were medium and any above .8 were large.

$3.40 - 3.92 = -.52$ divided by the pooled standard deviation $(1.61) = -.32$.

This is another time when we are concerned about the absolute value rather than the positive or negative direction of the outcome.

We see on the basis of this calculation that even if we had a statistically significant finding, we have a rather weak effect size. What does this mean? It means that the practical significance is low, as we have not explained a great deal of the variance between the two groups. You need to report the effect size only if you have a significant finding. See Example 6.5 for an example of reporting Cohen's *d*.

Once we have the effect size (Cohen's *d*), we have all of the information we need to properly analyze and report the findings from our study.

Reporting an Independent-Samples T-Test

Once you have all of the information that you need to report the independent-samples t-test, the next step is to create a research report from which the reader will be able to assess the use of the tests and interpretation of the test in any subsequent discussion. See Example 6.6 for a possible write-up of the data.

Writing Hints for Independent-Samples T-Tests

From Example 6.6 we see a few writing hints that may help you state your findings as directly as possible.

> **EXAMPLE 6.6 POSSIBLE WRITE-UP OF THE INDEPENDENT-SAMPLES T-TEST**
>
> In order to determine if differences existed in the sight-singing skills of students in a select choir and those in a general choir, the researcher administered a sight-signing test and compared the groups by using an independent-samples t-test. The researcher established the homogeneity of variance through the Levene test for equality of variance, $F = .69$, $p = .41$. Although the general choir performed slightly better, $M = 3.92$, $SD = 1.80$, than the select choir, $M = 3.40$, $SD = 1.38$, the researcher found no significant differences between the two groups' sight-singing scores, $t(48) = -1.15$, $p = .26$.

- Per American Psychological Association guidelines, use first-person pronouns for clarity, especially when describing actions you took, while keeping all of your actions in past tense.
- State your findings in active voice (i.e., the researcher did X or the participants did X).
- State your findings in past tense.
- When reporting statistical information, put the label of the statistic (e.g., F, t, p) in italics.
- Generally speaking, you do not need to go beyond the hundredths decimal space.
- Remember that those findings that are not significant are *not* insignificant and you should not use that language.
- Instead of reporting degrees of freedom parenthetically, you can also overtly state it ($t = -1.15$, $df = 48$, $p = .26$).
- Always overtly state the type of statistical procedure you are using (e.g., independent-samples t-test). Even if the well-informed reader can infer the type of test, it is helpful to overtly report it.

Final Thought on Independent-Samples T-Test

The independent-samples t-test is ubiquitous in educational research, as we often are trying to determine the effectiveness of one different pedagogy over another and can rarely randomly assign one. Thus we need a statistical procedure that can analyze the difference between two groups that are independent from each other (e.g., two different classes). Music education researchers and interpreters of research should know this procedure well and what each piece of information means in order to make some of the most elemental and meaningful decisions in our classrooms.

Independent-Samples T-Test Exercises

Given the second provided dataset, Independent Samples T-Test Example 2 ⊙, compute the most logical analysis needed to come to an informed conclusion. In this example the columns include

- Intonation Test Score (0–10 possible score)
- Group 1 = Brass Players, Group 2 = Woodwind Players

Conduct an analysis of the two groups' normal distribution in order to decide whether or not you should continue with the analysis and defend your choice by using empirical evidence. Once you have computed the most appropriate statistical test, write a report that includes the descriptive statistics necessary to understand the t-test as well as the effect size.

7 | Univariate Analysis of Variance (ANOVA)

Use of the Analysis of Variance

Researchers can use the univariate analysis of variance (ANOVA) when they need to compare the means of two or more groups or levels of a variable. Keep in the mind that if you have only two groups or levels of a variable (e.g., comparing wind players to percussionists in your band), the more appropriate test is the t-test. The most simple and parsimonious version of a statistical procedure is almost always the best.

As seen in Table 1.6 in Chapter 1, researchers using an ANOVA are seeking to answer a simple question: does a difference exist between two or more groups on one dependent variable?

For example, a music education researcher may be interested in exploring differences in students' sight-reading skills between students who are rising into high school from three different feeder schools. Because there are more than two groups, the t-test is no longer an option, so the researcher would need to employ an ANOVA. Another name for the univariate analysis of variance is a one-way ANOVA. Regardless of what it is called, the univariate ANOVA is a statistical procedure that focuses on a single independent variable as well as a dependent variable. You could think of the univariate ANOVA as an extension of the independent-samples t-test. Whereas the t-test could be used only with two groups or levels of a variable, an ANOVA can explore the differences between three or more groups or levels of a variable.

When music education researchers use the univariate ANOVA to examine the differences between three or more groups, the independent variable (*must be categorical*) is the different groups in the model and the dependent variable is a score (*must be continuous*) on some research instrument, test, or other measurement.

In Example 7.1 we can see that researchers used the univariate ANOVA to examine the differences in means among three groups that experienced different practice conditions. However, because the researchers found no significant differences, they did not report the actual results of the ANOVA.

> **EXAMPLE 7.1 UNIVARIATE ANOVA USED TO DETERMINE DIFFERENCES AMONG THREE GROUPS**
>
> The third research question concerned differences in student attitude among the three practice conditions. A one-way ANOVA on the attitude scores revealed no significant differences. The average responses ranged from 3.08 to 3.25 on a 4-point scale, indicating that students had an overall positive response to practice in all three practice conditions.
>
> ---
>
> Stambaugh, L. A., & Demorest, S. M. (2010). Effects of practice schedule on wind instrument performance: A preliminary application of a motor learning principle. *Update: Applications of Research in Music Education, 28*(2), 20–28.

> **EXAMPLE 7.2 UNIVARIATE ANOVA USED TO FIND SIGNIFICANT DIFFERENCES BETWEEN FOUR GROUPS**
>
> A one-way analysis of variance was used to assess the difference in mean scores across the four experimental conditions. Significant results were found, (F = 4.52, df = 3, p = .006 and reported in a Table 1). Means and standard deviation scores across conditions are contained in a Table 2. Tukey post hoc/multiple comparisons test was calculated to determine significance among the mean scores and reported in a Table 3 (see Figure 7.1). Significant results were found between spoken story with distraction condition (SD) and musical story with distraction condition (MD). No additional significant results were found among the remaining conditions.
>
> ---
>
> Wolfe, D. E., & Noguchi, L. K. (2009). The use of music with young children to improve sustained attention during a vigilance task in the presence of auditory distractions. *Journal of Music Therapy, 46*(1), 69–82.

It is unfortunately more common for researchers to report the actual outcome of an ANOVA when they find a significant difference. It is proper and useful to other researchers and profession as a whole to report the statistical outcome without regard to significance as well. As you can see in Example 7.2, researchers used the univariate ANOVA to examine the differences among four different treatment groups.

Post Hoc Tests for ANOVA

As one can see in Example 7.2, the researchers reported, in addition to the ANOVA statistics, the outcome of post hoc tests, which are those done after the fact (in this case, the fact is the ANOVA itself). It makes sense that if we are comparing only two groups and we find a significant difference between the means, then we can use the means of the two groups to know which group scored statistically higher or lower. What if, however, we compared the means of four groups, as in Example 7.2? The ANOVA results tell us only that somewhere in the data a difference in some means exists. We do not know, however, if there are one or more differences or between which groups the differences exist. We must therefore conduct post hoc tests in order to discover where the differences lie. When conducting a post hoc test in SPSS, we often get a readout that lets us compare each group within the ANOVA to every other group. To continue with the same study used in Example 7.2, note the set-up of the table in Figure 7.1.

In Figure 7.1, we can see that significant differences ($p < .05$) exist between the spoken, with distraction condition and music, with distraction condition ($p = .026$), as reported in Example 7.2. The researcher found no other significant differences. Note that one difference was close to reaching significance ($p = .052$), but the researchers did the ethical thing of not discussing it as approaching significance. They held true to the a priori decision to maintain a .05 significance level.

Assumptions of Univariate ANOVA

The univariate ANOVA has the same five major assumptions as the two t-tests discussed in previous chapters, except for the need for the independent variable to be bivariate. The ANOVA can handle more than two groups or levels in

Tucky Post Hoc/Multiple Comparisons Test

(I) COND	(J) COND	Mean difference (I–J)	Std. error	p
SND	SD	.72	1.927	.982
	MND	−4.47	1.957	.111
	MD	−4.90	1.875	.052
SD	SND	−.72	1.927	.982
	MND	−5.19	2.029	.059
	MD	−5.62*	1.949	.026
MND	SND	4.47	1.957	.111
	SD	5.19	2.029	.059
	MD	−.43	1.979	.996
MD	SND	4.90	1.875	.052
	SD	5.62*	1.949	.026
	MND	.43	1.979	.996

*The mean difference is significant at the .05 level.
Note. Abbreviations: SND = Spoken, no d--istraction, SD = Spoken, with distraction, MND = Music, no distraction, MD = Music, with distraction.

FIGURE 7.1

the categorical independent variables. The assumptions of the ANOVA, therefore, are the following:

1. The data collected for the dependent variable is continuous rather than categorical (e.g., scores on a musical playing test).
2. The data collected for the independent variable is categorical (e.g., what instrument a student plays, which ensemble a student is in).
3. Each observation is independent of any other observation.
4. The dependent variable is normally distributed (see Chapter 1).
5. The dependent variable has roughly the same variance in each group (variance is σ^2, as discussed in Chapter 1). Such consistency helps ensure that we are comparing apples to apples. This assumption is often referred to as homogeneity of variance.

Again, it is easier to argue for robustness against any violation of these assumptions if you have roughly equal numbers of participants or data points in each group and if there are more than 30 participants. The ANOVA is, however, even more robust against the violations than t-tests.

Meeting the Assumptions of the Univariate ANOVA

Meeting all of the assumptions of an ANOVA can be challenging. Music education researchers should strive, however, to conduct inferential statistical procedures only when it is appropriate. Otherwise, we are using poor information to inform our profession. In the case of the ANOVA, how to meet a few of the assumptions is a little less clear than in the t-tests. The dependent-variable data should be continuous, which is an easy assumption to meet as well and is considered long before data collection. Each observation should be independent from others. That is, the score of one participant should not be influenced by the score of any other participant. For example, a music education researcher should not give students a score for their entire instrumental section and then use that data as a dependent variable in an ANOVA.

When we get to assumptions 4 and 5, normal distribution and homogeneity of variance, meeting assumptions is a little more difficult. See Chapter 1 to review the methods for determining the normal distribution of a sample. You can use a histogram, or calculations of skewness and kurtosis, of your data to visually check the normal distribution. If, however, the groups were not normally distributed, it would be more appropriate to use the nonparametric version of the univariate ANOVA: the Kruskal–Wallis one-way analysis of variance (see the section "Conducting a Univariate ANOVA in SPSS"). As we did in previous chapters regarding t-tests, we can use the Levene test to examine homogeneity of variance as we did for the t-tests.

Research Design and the Univariate ANOVA

As discussed, the univariate ANOVA can be used to compare two or more groups or levels of a variable. So, a few of the possible research designs that utilize this type of inferential test include (O = an observation, X is a treatment of any given type):

(A)   O X_1 O

       O X_2 O

       O X_3 O

       O X_4 O

(B) X_1 O

      X_2 O

          O

(C)   O

     O

     O

Let's take a look at a few musical examples of what these designs might be in the area of music education.

(A) $O_{\text{diagnostic sight reading test}}$ $X_{\text{control group}}$ $O_{\text{final sight-reading test}}$

     $O_{\text{diagnostic sight reading test}}$ $X_{\text{1 minute of sight reading}}$ $O_{\text{final sight-reading test}}$

     $O_{\text{diagnostic sight reading test}}$ $X_{\text{5 minute of sight reading}}$ $O_{\text{final sight-reading test}}$

     $O_{\text{diagnostic sight reading test}}$ $X_{\text{10 minute of sight reading}}$ $O_{\text{final sight-reading test}}$

In the example design above, a music education researcher was interested in the impact of different lengths of sight-reading instruction on students' sight-reading achievement. First, the researcher administered a diagnostic sight-reading test to see where students' skills existed prior to any treatment. The first group received no sight-reading instruction (the control group). The second group received one minute of sight-reading instruction per class. The third group of students received 5 minutes of sight-reading instruction, while the fourth group received 10 minutes. The researcher could use an ANOVA to find out if any differences in mean gain score for the four different treatments. In this example, the students' mean gain score from first observation to second observation is the continuous dependent variable, while the categorical independent variable is group membership (i.e., members of first, second, third, and fourth treatments).

As you can imagine, there are countless possible alterations to this design. There could be only three groups, or even more than four groups. Not all groups have to have a treatment.

If the researcher uses an ANOVA in design B below, she can examine the differences between any number of groups without a pre-test by using the control group as the comparison point.

(B) $X_{instrumentalists}$ $O_{music\ preference\ test}$

$X_{vocalists}$ $O_{music\ preference\ test}$

$O_{music\ preference\ test}$

In this example (B), a researcher was interested in first-year college students' preferences of 20th-century art music examples and the impact that being in high school choir or an instrumental ensemble may have. The researcher would be able to use an ANOVA to compare the preference ratings (dependent variable) of the participants in each group (independent variable), including students who did not participate in a formal music ensemble during high school, for the sake of comparison.

In design C below, we see that no actual experimental design exists. Rather, researchers may use the univariate ANOVA to examine the differences between groups that are based upon individual-difference variables (those which were not manipulated or controlled by the researcher such as instrument family, level of education (bachelor's, master's, doctorate), etc.

(C) $O_{undergraduate\ music\ education\ student\ response\ to\ survey\ of\ teacher\ attributes}$

$O_{masters\ in\ music\ education\ student\ response\ to\ survey\ of\ teacher\ attributes}$

$O_{doctoral\ music\ education\ student\ response\ to\ survey\ of\ teacher\ attributes}$

In this example (C) a researcher could use an ANOVA to see if differences existed among three different groups on a survey of the attributes the students felt were important to be a successful music educator. The researcher is looking to see if those at different educational career stages have different beliefs as to what is most important.

This type of use of the ANOVA is most commonly found in descriptive studies. See Example 7.3.

The reporting of the actual ANOVA findings from Example 7.3 is seen in Figure 7.2 in a table format.

EXAMPLE 7.3 USE OF THE UNIVARIATE ANOVA TO EXAMINE INDIVIDUAL DIFFERENCES

Three ANOVAs were conducted to explore the impact of grade level, length of study, and instrument type on overall discomfort. The overall discomfort mean was used as the dependent variable, while grade (sixth, seventh, and eighth); the instrument participants played (violin, viola, cello, bass); and length of time they had studies (this year, two years, three years, four years, more than four years) were independent variables. As before, a Bonferroni adjustment was made (.05 ÷ 3 = .02) to diminish the threat of a Type I error. Neither which instrument the participant played nor how long he had played his instrument had a significant impact on overall discomfort. The grade a student was in, however, did have a significant impact on her overall discomfort. Because the ANOVA on the grade of the participants does not meet the assumption of equal variance, a Dunnett C post hoc test was conducted to discover where specific differences existed. The post hoc test revealed that significant ($p = .004$) differences exist between sixth-grade ($M = 1.56$) participants and participants in the seventh grade ($M = 1.25$), and between sixth grade and eighth grade ($M = 1.34$, $p = .012$) participants. No significant difference was found between the seventh and eighth grades. Sixth-grade participants reported significantly higher levels of discomfort than participants in the other two grades. However, the practical significance of this finding was limited (eta^2 of .076; see Figure 7.2).

Russell, J. A. (2006). The influence of warm-ups and other factors on the perceived physical discomfort of middle school string students. *Contributions to Music Education, 33*(2), 89–109.

ANOVA Tables on Impact of Instrument, Grade, Period of Study on Discomfort

	Levene Sig.	Sum of Squares	df	Mean Square	F	Sig.	Eta2
Instrument	.146	601.13	3	200.38	.62	.603	.012
Grade	.000	3851.88	2	1925.94	6.41	.002	.076
Period of Study	.000	3204.66	4	801.17	2.60	.038	.063

FIGURE 7.2

The Null Hypothesis for the Univariate ANOVA

The null hypothesis for the univariate ANOVA is that the means of all groups or levels of a variable will be equal (see Chapter 1 for a review of hypothesis statements). It could be stated as

$H_0: \mu_1 = \mu_2 = \mu_3 = \mu_4 = \mu_5$ etc.

For our musical example above regarding the impact of different lengths of sight-reading instruction, the null hypothesis would be that the means of the different treatment groups would be statistically the same.

$H_0: \mu_{\text{control group}} = \mu_{\text{1minute of sight reading instruction}} = \mu_{\text{5minutes of sight reading instruction}}$
$= \mu_{\text{10minutes of sight reading instruction}}$

The alternative hypothesis would be that there would be a difference between the mean of each of the groups or levels. It would be stated as

$H_a: \mu_1 \neq \mu_2 \neq \mu_3 \neq \mu_4 \neq \mu_5$ etc.

In our musical example, the alternative hypothesis could be written as

$H_a: \mu_{\text{control group}} \neq \mu_{\text{1minute of sight reading instruction}} \neq \mu_{\text{5minutes of sight reading instruction}}$
$\neq \mu_{\text{10minutes of sight reading instruction}}$

Finally, the directional hypothesis would state that the mean of one or more of the groups or levels of variable would not be equal and that one would be higher or lower than the other. This hypothesis would be stated as

$H_1: \mu_1 < \mu_2 < \mu_3 < \mu_4 < \mu_5$ etc.

In our musical example, the directional hypothesis might be listed as

$H_1: \mu_{\text{control group}} < \mu_{\text{1minute of sight reading instruction}} < \mu_{\text{5minutes of sight reading instruction}}$
$< \mu_{\text{10minutes of sight reading instruction}}$

Notice that in this directional hypothesis, the researcher is asserting that with more sight reading instruction comes greater achievement in sight reading. Although this is a logical assumption to make, a researcher should be wary of directional hypotheses unless they are supported by evidence from previous researchers.

Setting Up a Database for the Univariate ANOVA

The database for the univariate ANOVA will have *two* necessary columns. One column will have the dependent variable (some continuous variable) and the other column will have the independent variable. The independent

	Instrument	Performance
1	1	5
2	2	2
3	4	5
4	3	1
5	2	3
6	1	5
7	4	4
8	1	5
9	1	4
10	1	5
11	1	5
12	2	2
13	2	3
14	3	2
15	2	3

FIGURE 7.3

variable should look something like the grouping variable of the independent samples t-test, but can have more than two possible groupings. In Figure 7.3, we can see a database established to compare the scores of four different groups of instruments (strings, winds, brass, percussion), which have been assigned a number (i.e., strings = 1, winds = 2, brass = 3, percussion = 4) on their scores on a fictitious performance assessment (scaled 1–5).

Conducting a Univariate ANOVA in SPSS

For this section, please use the provided example database (Univariate ANOVA Example 1 ▶ in order to go through the process while reading. This exercise will give you the guided experience of a step-by-step analysis so that you can conduct the follow-up analysis on the additional dataset on your own.

You have most likely already encountered the menu-item way when you explored the independent-samples and dependent-samples t-tests. In order to run the univariate ANOVA, follow these steps:

1. Click **Analyze**.
2. Click **Compare means**.
3. Click **One-Way ANOVA**.
4. Enter the dependent variable in the dependent list. Note that SPSS will allow you to run multiple one-way ANOVAs through this one screen by entering multiple dependent variables, but only one independent variable (factor).
5. Enter the independent variable in the **Find Factor(s)** space.

6. Click on **Options**.
7. In the **Options** screen make sure to select **Descriptives** and **Homogeneity of variance test**.
8. Make sure that the missing values are being excluded analysis by analysis (this step will make sure SPSS does not reject any more data than it must).
9. Click **Continue**.
10. Click **OK**.

Once you have clicked **OK**, you should get three different boxes of information from SPSS: the descriptive statistics, homogeneity of variance test, and the **ANOVA** table. Let's look at the dataset given earlier in Univariate ANOVA Example 1 ▶. In this example, the researcher sorted music students into four groups based upon the participants' primary instrument (1 = strings, 2 = woodwinds, 3 = brass, 4 = percussion) and gave each student a performance test in which a student could score from 1 to 5. To run a univariate ANOVA with this data, we can follow the steps above and get the following information from the three boxes that SPSS provides.

In the first box (see Figure 7.4), we simply see the descriptive statistics of each of the four groups as well as the aggregated total.

From Figure 7.4, we can see that 23 string students, 22 woodwind students, 21 brass students, and 20 percussion students ($N = 86$) participated in the study. The mean of each group follows in the next column. We see that the string participants' ($n = 23$) mean score was 4.52 with a standard deviation of .73). For now, these are the three columns in which we are most interested.

The next box (see Figure 7.5) is the Levene test for homogeneity of variance. Remember that, in this instance, *you do not want to have a significant difference*. You are looking for a significance level that is .05 or greater. A number smaller than .05 means that your two groups do not have sufficiently similar variance to meet one of the fundamental assumptions of an ANOVA (homogeneity of variance). We can see from Figure 7.5 that we did indeed meet the assumption.

Descriptives

Performance

	N	Mean	Std. Deviation	Std. Error	95% Confidence Interval for Mean		Minimum	Maximum
					Lower Bound	Upper Bound		
1	23	4.52	.730	.152	4.21	4.84	2	5
2	22	2.50	.512	.109	2.27	2.73	2	3
3	21	1.67	.658	.144	1.37	1.97	1	3
4	20	4.30	.470	.105	4.08	4.52	4	5
Total	86	3.26	1.348	.145	2.97	3.54	1	5

FIGURE 7.4

Test of Homogeneity of Variances
Performance

Levene Statistic	df1	df2	Sig.
1.504	3	82	.220

FIGURE 7.5

ANOVA
Performance

	Sum of Squares	df	Mean Square	F	Sig.
Between Groups	124.266	3	41.422	112.823	.000
Within Groups	30.106	85	.367		
Total	154.372	85			

FIGURE 7.6

As the significance level (.220) is greater than .05, we can assume that the variance between the groups (notice that the first degrees of freedom is calculated by groups ($N - 1$) and the second degrees of freedom is calculated based upon participants minus the number of groups ($N - 4$) is similar enough to meet the assumption of homogeneity of variance.

The third box created by SPSS is the actual ANOVA table (see Figure 7.6). In this box we can see the sum of squares between groups, within groups, and total, as well as the degrees of freedom, mean square, F statistic, and significance level.

Understanding the ANOVA Table

It can be very easy for a researcher to simply glean the information required from the ANOVA table to write a good research report. However, the more we understand what the numbers mean, the better we may become at interpreting the data and making more meaningful inferences from our work. To that end, in this section of the chapter, we will take a look at how the data in the ANOVA table is calculated.

There are *five* pieces of information that we want to try to understand better in this table: between group sum of squares, within group sum of squares, total sum of squares, mean square within group, and mean square between groups. The first three pieces of information have to do with types of sum of squares calculated in the ANOVA. Try not to get bogged down by the phrase sum of squares, which means exactly what it states. A sum of squares is simply adding up the various scores that have been squared. Then we should try to understand how we achieved the mean square between groups and mean square within groups. The mean squares are really versions of average variation of scores found both within a group and between groups. Below is an example of how to calculate each of these pieces of information. This process can help the reader better understand what information actually goes into the data to improve understanding. They are discussed in the order that they are most easily calculated.

For review, consider the following:

- Between groups = comparing the data between each group or level of the independent variable, for example, comparing the variance found in instrumentalists' scores to that of the variance found in vocalists' scores.
- Within groups = comparing the variance found within all of the participants of the study with no concern for what group or level of the independent variable to which they belong.

1. *Between groups sum of squares*—this step shows the relative variance between groups by comparing the mean of each group to the overall mean. You can calculate the between-group sum of squares by subtracting the mean of each group from the total mean, then squaring each one prior to adding it all up. Let's check our sum of square by calculating our own with information from our musical database example. In Figure 7.4, we see that the overall mean is 3.26. We then take the mean of each group and subtract the overall mean from each one.

Strings = 4.52 − 3.26 = 1.26
Woodwind = 2.50 −3.26 = −.76
Brass = 1.67 − 3.26 = −1.59
Percussion = 4.30 −3.26 = 1.04

We then square each outcome (this step makes us not worry about any negative number, as the square of a negative number is always positive) and multiply it by the number of participants in each group:

Strings ($n = 23$) = 4.52 − 3.26 = 1.26 × 1.26 = 1.59 × 23 = 36.57
Woodwind ($n = 22$) = 2.50 − 3.26 = −.76 × −.76 = .58 × 22 = 12.76
Brass ($n = 21$) = 1.67 − 3.26 = −1.59 × 1.59 = 2.53 × 21 = 53.13
Percussion ($n = 20$) = 4.30 − 3.26 = 1.04 × 1.04 = 1.08 × 20 = 21.60

Once we have summed these four numbers, we get 124.06, or the between-group sum of squares. Note that the difference between our calculation and that offered by SPSS (124.266) is due to rounding error, as we rounded each number up to the hundredth. It is easy to see how these mathematical steps are attempting to take into account the number of participants in each group and the mean score of each group and compare it to the most accurate overall score (the overall mean). In using this algorithm, the researcher is trying to create a more objective comparison of the differences between separate groups while still taking into account each individual group size and score, as well as the aggregated mean.

The next step toward computing the actual F statistic is to calculate the mean square between groups.

2. *Mean square between groups*—in order to calculate the mean square between groups, take the sum of squares between groups and divide it by the degrees of freedom between groups (the number of groups in the analysis minus 1) in the analysis. In our example, it would be 124.06 ÷ 3, or 41.35 (again the slight difference in number from the one found in the SPSS output is due to rounding error). Why not just divide by the number of actual groups? That process would make the mean square smaller, meaning that it would be more difficult to find significance leading to a possible Type I error (refer to Chapter 1 for a refresher on degrees of freedom). If we had data for every person in the population, we could divide by the groups. As we are working with only a sample, we allow for more freedom (hence the name) in the mean difference.

Next, we need to understand the sum of squares within the groups.

3. *Sum of squares within the groups*—this number helps us understand the variation within the groups in the analysis. In order to calculate the sum of squares within groups we find the degrees of freedom for each group ($n - 1$) and then multiply that by the squared standard deviation of each individual group. We can get the standard deviation for each group from the descriptive readout from SPSS.

Strings ($n = 23$) = 23 − 1 = 22, SD^2 = .73 × .73 = .53
Woodwind ($n = 22$) = 22 − 1 = 21, SD^2 = .51 × .51 = .26
Brass ($n = 21$) = 21 − 1 = 20, SD^2 = .66 × .66 = .44
Percussion ($n = 20$) = 20 − 1 = 19, SD^2 = .47 × .47 = .22

We then take the degrees of freedom from each group and multiply it by the squared standard deviations of each group:

Strings: 22 × .53 = 11.66
Woodwinds: 21 × .26 = 5.46
Brass: 20 × .44 = 8.80
Percussion: 19 × .22 = 4.18

We add it all up and get 30.10. The SPSS printout offered 30.106. Again, the minimal difference is due to rounding error.

4. *Mean square within groups*—in order to calculate the within mean square, we need to compute the degrees of freedom within groups. To do so, subtract the number of groups from the overall number of participants. In our example, the degrees of freedom within groups are 82 (86 − 4).

Then to calculate the within mean square, we divide the within sum of squares by the degrees of freedom within groups.

$$30.10 \div 82 = .37$$

The within mean square is .37.

5. *Total sum of squares*—the last piece of information is the total sum of squares, which is the easiest to calculate. It is simply adding the sum of squares between groups to the sum of squares within groups. This calculation will help the researcher compute an effect size indicator for the ANOVA. More on that point follows.

So what does all of this mean? It may seem a little like mathematical minutiae to worry about the sum of squares within and between groups or even the mean of the sum of squares within and between groups. However, with these pieces of information, you can easily calculate the actual F statistic in the ANOVA. In short, the F statistic is just a ratio (also known as a comparison) of the difference of the between-group variance and the within-group variance and is simply computed by dividing the between-group mean square by the within-group mean square.

$$F = \text{Between Group Mean Square} \div \text{Within Group Mean Square}$$

In our example, $F = 41.35 \div .37 = 111.75$ (again, slightly different due to rounding error).

On the basis of this example, what can we tell about the relationship between the between mean score and the within mean score? The greater the difference between the two, the greater F is going to be. Therefore, a larger F statistic is likely to lead to a significant finding. F is simply a figure that tells us how different the groups were when you were taking into account the variance between groups as well as the variance within groups while also taking into account the number of groups and the number of individual participants. You can use the SPSS printout to find the significance level or take a look at a traditional F Table.

Do we have enough information to know where the difference is? Not quite yet. We did find a significant difference between groups. However, we have no idea if the difference is between the string players and the brass players, or between the woodwinds and brass players, or any combination of groups. In order to examine where actual differences exist, we need to conduct some post hoc tests.

The first step in conducting these tests is to decide if you have equal variance in your groups or not. Check the test of homogeneity of variance. This step is important, as it will inform what type of post hoc test we should run. In this instance we can assume equal variance, as we did not violate the assumption of homogeneity of variance as evidenced in the Levene test.

To run a post hoc test in SPSS, follow these steps:

1. *Click* **Analyze**.
2. *Click* **Compare means**.
3. *Click* **Run One-Way ANOVA**.

4. *Enter the dependent variable in the dependent list. Note that SPSS will allow you to run multiple one-way ANOVAs through this one screen by entering multiple dependent variables, but only one independent variable (factor).*
5. *Enter the independent variable in the* **Find Factor(s)** *space*

N.B.: Steps 1 through 5 should be the exact same as your original analysis steps.

6. *Click* **Post Hoc**.
7. *You will see two boxes, one with a list of tests to use with equal variances assumed and a smaller box with tests to choose when you cannot assume equal variance. Since we met the assumption of homogeneity, we will select one from the box of tests that assumer equal variance. I strongly suggest you employ the Scheffé post hoc test as it is one of the most flexible as well as conservative tests. Select* **Scheffé**.
8. *Click* **Continue**.
9. *Click* **OK**.

N.B. If you cannot assume equality of variance, but wish to persist with the post hoc tests, I suggest the **Games-Howell** post hoc test.

SPSS will create the exact same readout as you received the first time you conducted your analysis. You may ask, then, why not just do it when you first run your analysis? The answers are that you may not need to do a post hoc test at all if you find that there is not significance found overall, and you will not know if you can assume equal variance until you complete the Levene test. Therefore, you would not know what type of post hoc test to complete. Why make your first step into your analysis more complicated by creating extra readouts you may not need?

In addition to the same readout as before, SPSS will also create the post hoc comparisons (see Figure 7.7).

From this table (using the significance column) we can see that the actual differences existed between groups 1 and 2 (i.e., strings and woodwinds), 1 and 3 (i.e., strings and brass), 2 and 3 (i.e., woodwinds and brass), 2 and 4 (woodwinds and percussion), and 3 and 4 (brass and percussion). The only two groups that were not significantly different from each other were groups 1 and 4 (i.e., strings and percussion).

In order to find out the direction of the differences, we need to go all the way back to the descriptive data from the ANOVA found in the SPSS output. Using these descriptive statistics, we find that string players scored a mean of 4.5, while the woodwind players scored a mean of 2.5. The brass players had a mean score of 1.6 and the percussionists scored a mean of 4.3. We can now interpret the findings from the post hoc test:

Multiple Comparisons

Dependent Variable: Performance
Scheffe

(I) Instrument	(J) Instrument	Mean Difference(I−J)	Std. Error	Sig.	95% Confidence Interval	
					Lower Bound	Upper Bound
1	2	2.022*	.181	.000	1.51	2.54
	3	2.855*	.183	.000	2.33	3.38
	4	.222	.185	.699	−.31	.75
2	1	−2.022	.181	.000	−2.54	−1.51
	3	.833	.185	.000	.31	1.36
	4	−1.800	.187	.000	−2.33	−1.27
3	1	−2.855*	.183	.000	−3.38	−2.33
	2	−.833*	.185	.000	−1.36	−.31
	4	−2.633*	.189	.000	−3.17	−2.09
4	1	−.222	.185	.699	−.75	.31
	2	1.800*	.187	.000	1.27	2.33
	3	2.633*	.189	.000	2.09	3.17

*The mean differences is significant at the 0.05 level.

FIGURE 7.7

- String players scored significantly higher than woodwind players and brass players.
- Woodwind players scored significantly higher than brass players
- Percussionists scored significantly higher than woodwind players and brass players.
- No statistically significant difference existed between string-player and percussion-player scores.

In Example 7.4, we can see an in-text reporting of a post hoc test.

If you have a larger number of significant post hoc findings, you may find it easier to report them in table form.

Is this enough information yet? Still not quite yet. We still have not talked about effect size. As with the t-tests, we need to know how much variance is being accounted for before we can discuss the actual *practical significance* of any finding. If you had no statistically significant findings, yes, you are done and can adequately report your findings. However, if your omnibus test (a generic name given to any test that examines statistical differences and requires follow-up analyses such as post hoc tests; in this case, the ANOVA is the omnibus test) was significant, you need to let the reader know about the effect size so that the findings may be better interpreted.

EXAMPLE 7.4 IN-TEXT REPORTING OF POST HOC TESTS

There were also significant differences by grade level for all participants (boys/girls, Hispanic/African American) to several of the attitude statements. In response to the statement "I like to sing in music class," the third graders were more positive than fifth graders, $F\,(2, 164) = 4.41$. The Scheffé post hoc test indicated that third graders were also significantly more positive than fifth graders in regard to the statement "I think everyone should sing, not just singers on radio, TV, and records," $F\,(2, 164) = 3.77$. It is interesting to note that fifth graders were sometimes more positive than the younger participants toward some of the statements and, although there were some significant differences by grade level, the lowest mean response was still 2.42 on the 5-point scale.

Siebenaler, D. (2008). Children's attitudes toward singing and song recordings related to gender, ethnicity, and age. *Update: Applications of Research in Music Education, 27*(1), 49–49–56.

The most common reporting of effect size for ANOVAs is eta squared (η^2). Computing η^2 is actually one of the easiest things you will do by hand while calculating anything within the ANOVA.

Simply divide the between-group sum of squares (SS) by the total sum of squares. Both of these pieces of information are provided in the ANOVA table in SPSS.

$$\eta^2 = Between\,Group\,SS \div Total\,SS$$

In our musical example from our practice database, $\eta^2 = 124.266 \div 154.372 = .80$.

In this example, our grouping (primary instrument) accounted for 80% of the variance between groups. This percentage would argue for strong practical significance of the findings. Generally speaking, anything below .40 is weak, .41–.69 is moderate, and .70 and above is a strong η^2. See Example 7.5.

Reporting a Univariate ANOVA

Once you have conducted the ANOVA by using the provided data set for this chapter, the final step is to create a research report that could be offered in a journal setting or dissertation writing. For readers not currently conducting their own research, looking at how these tests go from inception to reporting can give you a keen eye in determining the trustworthiness of the research report and the data contained within.

EXAMPLE 7.5 EXAMPLE REPORTING OF ETA SQUARED

Calculations of eta squared from an ANOVA summary table showed that singing mode accounted for 20% of the total variance in ratings over all 102 judges, 58% in choral conductors, 7% in voice teachers, and 4% in nonvocal musicians. Seating arrangement accounted for 7% of the variance overall, 4% in choral conductors, 17% in voice teachers, and 3% in nonvocal musicians.

Ekholm, E. (2000). The effect of singing mode and seating arrangement on choral blend and overall choral sound. *Journal of Research in Music Education, 48*(2), 123–135.

EXAMPLE 7.6 POSSIBLE UNIVARIATE ANOVA REPORT OF THE EXAMPLE DATA SET

In order to examine the differences between the means of different groups of instrumental students ($N = 86$) on a standard performance test, the researcher conducted a one-way ANOVA in which the independent variable was the instrument grouping—strings ($n = 23$), woodwinds ($n = 22$), brass ($n = 21$), and percussion ($n = 20$)—and the dependent variable was the score on the performance test. The researcher found a significant difference between groups ($F = 112.82$, $df = 3$, $p < .001$) as well as a strong η^2 (.80), suggesting practical as well as statistical significance. The researcher conducted a Sheffe post hoc test to determine between which groups actual differences existed. String players ($M = 4.52$, $SD = .73$) scored significantly higher on the performance test than woodwind players ($M = 2.50$, $SD = .51$) and brass players ($M = 1.67$, $SD = .11$). Percussionists ($M = 4.30$, $SD = .47$) also scored significantly higher than woodwind or brass players. Woodwind players scored significantly higher than brass players.

Given the grouping of the participants in our running music example in this chapter, one possible report of the data may look like Example 7.6.

Writing Hints for Univariate ANOVA

From the report in Example 7.6, we see some hints that might help you state your findings as clearly as possible while providing all of the necessary information.

- Make sure that you provide the following in your report:
  - N
  - n

- o *F*
- o *df*
- o *p*
- o η^2
- o Discussion of practical and statistical significance
- o Means and standard deviations of all groups
- o What type of ANOVA and post hoc tests you used and why
- Do not repeat the means of groups more than once. One you have reported it, you do not need to do so again.
- Another way to report the means and standard deviations of the groups is in a table.
- Per American Psychological Association guidelines, use first-person pronouns for clarity, especially when describing actions you took, while keeping all of your actions in past tense.
- Use the active voice.
- Avoid personification (e.g., "data indicated").
- Make sure all statistical indicators are in italics.
- Keep your actions as researcher in the past tense.

Univariate ANOVA Exercises

Given the second provided dataset, Univariate ANOVA Example 2 ▶, compute the most logical analysis. In this example, the two columns are the following:

- Educational Background (group 1 = Bachelor of Arts, group 2 = Bachelor of Music, group 3 = Bachelor of Music Education, group 4 = Bachelor of Science).
- Score on New Music Teacher Preparation Scale (1 = not prepared through 10 = very prepared.

In this example dataset, a researcher was interested in whether or not the type of degree a music education student received, and any differences in training based on that different degree classification that may exist, impacted participants' scores on a rating of new music teacher preparation. Was one group more prepared to teach than any of the others? What differences, if any, existed between the four groups of participants? These are the questions that can be answered when a researcher does a thorough job of collecting good information and employs the most appropriate statistical procedure in order to infer the best information possible.

Once you have computed the most appropriate statistical test, write a report that includes any descriptive statistics necessary to understand the ANOVA as well as the test itself and, finally, effect size.

8 | Factorial Analysis of Variance (ANOVA)

Use of the Factorial ANOVA

What can a researcher use to compare the differences between groups on the basis of more than one independent variable and still only one dependent variable? As long as the independent variables are between groups, rather than within groups (independent rather dependent), they can use a factorial ANOVA. A factorial ANOVA is an ANOVA that can help a researcher decide on the basis of more than one between group independent variable or even the *interactions* between multiple independent variables whether or not two groups are different.

When music education researchers use a factorial ANOVA, they are attempting to answer two research questions. Does a difference exist between two or more groups on more than one independent variable and one dependent variable? And does an interaction exist between the independent variables?

For example, a music education researcher might be interested in examining the outcome of in-service music educators' scores on an ensemble conducting test. The researcher may be most interested to see if self-identified gender and highest earned degree had any impact on the score. In this case, it would be appropriate for the researcher to use a factorial ANOVA, as the dependent variable is continuous (score on a conducting assessment) and the two independent variables are categorical (gender and highest earned degree).

At this point, you might be asking, why not just use two ANOVAs instead of a factorial ANOVA? The answer is really twofold. The first is that including both independent variables in which the researcher is interested in the same model diminishes the threat of a Type I error, as she would be making fewer observations. The second is that we might miss some important information if we analyzed the two independent variables separately. For example, we may find that it is really women with master's degrees who are above the rest of the participants in the study, a factor that may have been missed if we had taken a look at the data separately.

Factorial ANOVAs are usually accompanied by a grouping of numbers (e.g., 2 × 2 factorial ANOVA, 2 × 3 factorial ANOVA, etc.). The numbers tell the reader a great deal about the analysis. How many numbers there are tell the reader how many different independent variables the researcher included in the analysis. So a 2 × 3 ANOVA has two independent variables, while a 2 × 3 × 2 ANOVA has three independent variables. The actual numbers listed tells the reader how many levels the independent variable has. A 2 × 3 ANOVA has two independent variables, one with two levels, and one with three levels. An example of two levels may be gender (male or female) and an example of three levels may be school level taught (elementary school, middle school, high school). In our example above regarding the outcome of a conducting test, we would have a 2 × 3 factorial ANOVA (assuming participants identified as only one of two genders), as we would have two levels of gender and three levels of highest earned degree (bachelor's, master's, doctoral).

Some more examples to illustrate the point:

1. A 2 × 2 × 2 ANOVA has three independent variables, of which all have two levels (e.g., gender, certified to teach or not, instrumentalist or singer).
2. A 2 × 3 × 4 ANOVA has three independent variables, one of which has two levels (e.g., certified or not), one has three levels (teach in a rural, suburban, or urban school), and one has four levels (no music education degree, music education bachelor's degree, music education master's degree, music education doctorate).

The goal of a factorial ANOVA is to examine the *main effects* of each independent variable on the dependent variable as well as any *interactions* between the independent variables. Let's take the last example to clarify. In the 2 × 3 × 4 ANOVA, we would use the factorial ANOVA to tell us if any main effects existed; that is, did certification status, teaching setting, or highest music education degree earned impact the dependent variable? However, that would be the same as just running three individual one-way ANOVAs. The factorial ANOVA can also tell researchers whether or not any interactions existed between the independent variables; that is, were those who were certified but had not earned a music education degree significantly related to the dependent variable? That would be an example of an interaction between the certification-status independent variable and the highest earned music-education-degree variable.

As stated above, the factorial ANOVA distinguishes itself from the one-way ANOVA in that it can examine the interactions between multiple independent variables.

In Example 8.1, we can see how researchers used a 2 × 2 ANOVA to examine the differences between the treatment and control groups while also considering the grade level of the participants.

> **EXAMPLE 8.1** REPORTING OF A 2 × 2 FACTORIAL ANOVA
>
> To determine whether the use of a root melody accompaniment in song instruction has an effect on kindergarten and first-grade children's tonal improvisations, a two-way analysis of variance (treatment by grade) was conducted ($p \leq .05$). Results for the tonal improvisation scores (see Table 4) indicated no significant interactions. However, there were statistically significant differences for the main effects of treatment and grade level. Children in the experimental group received significantly higher improvisation ratings than children in the control group. The mean scores of the first-grade children were significantly higher than those of children in kindergarten.
>
> ---
>
> Guilbault, D. M. (2004). The effect of harmonic accompaniment on the tonal achievement and tonal improvisations of children in kindergarten and first grade. *Journal of Research in Music Education, 52*(1), 64–76.

In Example 8.1, the researcher did not find a significant interaction effect, but did find significant main effect differences. If, however, a researcher were to find significant interaction effects, it might be appropriate to report the interaction as well as main effects, but the researcher should base his conclusions on the interaction effect, as they are the most parsimonious interpretations to make (see Example 8.2).

In Example 8.2, the authors reported that they conducted a 2 × 2 ANOVA, the main effects (which look just like the reporting of a one-way ANOVA), as well as the interaction effect. Notice that for each *F* test, the authors provided the actual *F* statistic, the degrees of freedom between (the first number in parentheses), the degrees of freedom between (the second number in parentheses), the significance, and an effect size (eta). In addition to this information, the authors elected to provide the confidence intervals for each analysis. Although somewhat informative, it is not usually necessary to report the confidence intervals for the tests.

Post Hoc Tests for Factorial ANOVA

As with the one-way ANOVA, it may be necessary to conduct post hoc tests to see where any significant differences exist. If an independent variable has only two levels (e.g., gender) there is no need for a post hoc test. If, however, an independent variable as three or more levels, the researcher will need to conduct a post hoc test to see where the differences actually exist. In Example 8.3, researchers report the post hoc test of a 4 × 2 × 2 ANOVA.

EXAMPLE 8.2 REPORTING OF SIGNIFICANT INTERACTION EFFECT IN A 2 × 2 ANOVA

To explore singing accuracy in the pitch-matching task, a 2 × 2 ANOVA on the percentage of tuning errors was conducted with a diagnosis group and music training as between-subject factors. This analysis revealed a significant main effect for the diagnosis group, with results in favor of the control group, F (1, 26) = 4.96, $p < .05$, $\eta^2 = .07$, confidence interval (CI) .95 = 1.48 (lower) and 36.93 (upper). Participants with music training presented a lower percentage of tuning errors than the untrained participants both in the Williams syndrome and the control groups, F (1, 26) = 41.51, $p < .001$, $\eta = .56$, CI .95 = 37.84 (lower) and 73.30, (upper). The interaction effect of diagnosis group and music training was also significant, F (1, 26) = 4.50, $p < .05$, $\eta = .06$, CI .95 = –72.05 (lower) and –1.13 (upper).

Martínez-Castilla, P., & Sotilla, M. (2008). Singing abilities in Williams syndrome. *Music Perception, 25*(5), 449–469.

EXAMPLE 8.3 POST HOC REPORTING OF A FACTORIAL ANOVA

A 4 × 2 × 2 ANOVA) showed a main effect of intended emotion, F (3, 90) = 20.43, MSE = 1.84, $p < .001$, and post hoc comparisons showed that, collapsed across the presence or absence of lyrics, the participants gave higher ratings for the happy music than the sad, calm, and angry music, t's (126) = 5.95, 3.92, 7.09, respectively, all p's < .05. The ratings for the sad, calm, and angry music were not different from one another, all p's > .10.

Peynircioglu, Z. F., & Ali, S. O. (2006). Songs and emotions: Are lyrics and melodies equal partners? *Psychology of Music, 34*(4), 511–534.

Reporting Factorial Analysis of Covariance in Table Format

In order to save some space, if there is a need for it, researchers can report factorial ANOVA findings in table format. Often, the table looks just like a one-way ANOVA table (see Chapter 7). As seen in Figure 8.1, the author has reported a factorial ANOVA in table form. In the rows, the researcher has listed the pertinent information for the main effects of treatment and grade, as well as a row that shows the interaction of the two independent variables. In the columns, the author has reported the degrees of freedom, sum of squares (see Chapter 7), mean square, F, and the significance level). Missing is the report of effect size.

ANOVA Table of Improvisation Scores

	df	SS	MS	F	p
Treatment	1	11.71	11.17	6.08	.01*
Grade	1	10.45	10.45	5.43	.02*
Treatment by Grade	1	0.12	0.12	0.06	.80
Residual	32	254.23	1.93		

$p \leq .05$.

FIGURE 8.1

Assumptions of the Factorial ANOVA

The factorial ANOVA has basically the same assumptions as the one-way ANOVA discussed in Chapter 7. It is different, however, in that the factorial ANOVA can handle more than one independent variable and that all variables need to be between-group comparisons (not within group).

1. The data collected for the dependent variable is continuous rather than categorical.
2. Each observation is independent of any other observation.
3. The dependent variable is normally distributed (see Chapter 1).
4. The dependent variable has roughly the same variance in each group (variance is σ^2), a consistency that helps ensure that we are comparing apples to apples. This assumption is often referred to as homogeneity of variance.

Meeting the Assumptions of the Factorial ANOVA

As discussed in the previous chapters, meeting all of the assumptions of a factorial ANOVA is challenging. As discussed in Chapter 6, researchers should use only inferential statistical procedures when it is appropriate and the data best meets the assumptions of each test. In the case of the factorial ANOVA, meeting a few of the assumptions is relatively easy. The dependent-variable data should be continuous, which is an easy assumption to meet as well and is considered long before data collection. Each observation should be independent from others. That is, the score of one participant should not be influenced by the score of any other participant (don't give students a score for their entire instrumental section and then use that data as a dependent variable in a factorial ANOVA, or any kind of parametric ANOVA). As with the one-way ANOVA, when we get to assumptions 3 and 4, normal distribution and homogeneity of variance, the ease of meeting assumptions wanes. Remember, you can use a histogram or calculations of skewness and kurtosis of your data to visually check the normal distribution. As with all of the comparison of means discussed so far, we will use the Levene test to examine homogeneity of variance.

Research Design and the Factorial Analysis of Covariance

The factorial ANOVA can be used in any research design in which the ANOVA is employed. It can, however, be used to create a slightly more complex analysis, as it can examine the impact of more than one categorical independent variable on a single dependent variable. Regardless of the design used, it is important to remember that a factorial ANOVA is still a univariate (one dependent variable) statistical procedure. If your research design has more than one dependent variable, the factorial ANOVA is not an appropriate test.

As the use of the factorial ANOVA is relatively rare in purely experimental designs, the most common design of the factorial ANOVA is usually in a descriptive study. In such cases it can be challenging to look at a design in a traditional X and O style. The design of a factorial ANOVA is one that can look many different ways, depending on the number of independent variables and the number of levels in each. In our example from the beginning of the chapter, we have a 2 × 3 factorial ANOVA, which could be examined visually in the form of a table (see Figure 8.2).

As we can see in this figure, the researcher is interested in the main effects of both gender and highest degree earned, but is also interested in the interaction between these two independent variables. For instance, do men with higher degrees score differently from how women with higher degrees do? There are several questions to be answered by using this factorial design.

The Null Hypotheses for the Factorial ANOVA

Stating the null hypotheses for a factorial ANOVA is somewhat different from statements in tests discussed earlier in this book. First, the number of hypotheses depends upon the number of independent variables included in

		Independent Variable of Level of Degree with 3 levels		
		Bachelors Degree	Masters Degree	Doctoral Degree
Independent Variable of Self-Identified Gender with 2 Levels	Women	Mean Conducting Score	Mean Conducting Score	Mean Conducting Score
	Men	Mean Conducting Score	Mean Conducting Score	Mean Conducting Score

FIGURE 8.2

the factorial ANOVA. In a 2 × 2 factorial ANOVA, for instance, the researcher should state three hypotheses.

H_0: $\mu_{\text{independent variable 1 level 1}} = \mu_{\text{independent variable 1 level 2}}$
H_0: $\mu_{\text{independent variable 2 level 1}} = \mu_{\text{independent variable 2 level 2}}$

Also,

H_0: no interaction between independent variable 1 and 2 is present

In this example, the first two hypotheses are the main-effect hypotheses, while the third hypothesis is the interaction-effect hypothesis.

If, however, the researcher employed a 2 × 2 × 3 factorial ANOVA, the four null hypotheses would be

H_0: $\mu_{\text{independent variable 1 level 1}} = \mu_{\text{independent variable 1 level 2}}$
H_0: $\mu_{\text{independent variable 2 level 1}} = \mu_{\text{independent variable 2 level 2}}$
H_0: $\mu_{\text{independent variable 3 level 1}} = \mu_{\text{independent variable 3 level 2}} = \mu_{\text{independent variable 3 level 3}}$
H_0: no interaction between independent variable 1 and 2 and/or 3 is present

In our musical example of a 2 × 3 factorial ANOVA with gender and highest earned degree serving as independent variables and a conducting score serving as a dependent variable, our null hypotheses would be

H_0: $\mu_{\text{men}} = \mu_{\text{women}}$
H_0: $\mu_{\text{bachelors}} = \mu_{\text{masters}} = \mu_{\text{doctoral}}$
H_0: no interaction between independent variable 1 and 2 is present.

The alternative hypotheses for the 2 × 2 ANOVA would be

H_0: $\mu_{\text{independent variable 1 level 1}} \neq \mu_{\text{independent variable 1 level 2}}$
H_0: $\mu_{\text{independent variable 2 level 1}} \neq \mu_{\text{independent variable 2 level 2}}$
H_0: An interaction between independent variable 1 and 2 is present.

The alternative hypotheses for the 2 × 2 × 3 ANOVA would be

H_0: $\mu_{\text{independent variable 1 level 1}} \neq \mu_{\text{independent variable 1 level 2}}$
H_0: $\mu_{\text{independent variable 2 level 1}} \neq \mu_{\text{independent variable 2 level 2}}$
H_0: $\mu_{\text{independent variable 3 level 1}} \neq \mu_{\text{independent variable 3 level 2}} \neq \mu_{\text{independent variable 3 level 3}}$
H_0: An interaction between independent variable 1 and 2 and/or 3 is present

The alternative hypothesis for our musical example would be

$H_0 : \mu_{\text{men}} \neq \mu_{\text{women}}$
$H_0 : \mu_{\text{bachelors}} \neq \mu_{\text{masters}} \neq \mu_{\text{doctoral}}$
H_0: An interaction between independent variable 1 and 2 is present

Setting Up a Database for the Factorial ANOVA

The database for the factorial ANOVA will have at least three necessary columns. One column will have the dependent variable (a continuous variable) and the other columns will have the independent variables (both being between-group comparisons). The independent variable should look something like the grouping variable of the independent-sample t-test but, like the one-way ANOVA, can have more than two possible groupings. In Figure 8.3, we can see a database established to compare the conducting scores of both genders as identified by the participants (independent variable 1 where 0 = male and 1 = female) as well as the second independent variable, highest earned music education degree (1 = no music education degree, 2 = bachelor's in music education, 3 = master's in music education, 4 = doctorate in music education). Please note that this data differs from the running example discussed so far, as there is an additional level in this data for highest degree to account for the possibility of having no degree in music education. Hence, this design would be a 2 × 4 factorial ANOVA.

Conducting a Factorial ANOVA in SPSS

For this section, please use the provided example database, Factorial ANOVA Example 1 ▶, in order to complete the process while reading. This exercise will help you conduct the test step by step and learn how to complete future

	Gender	HighestMEdegree	ConductingOutcome
1	0	1	3
2	0	1	4
3	0	1	5
4	0	1	3
5	0	1	5
6	0	1	4
7	0	1	3
8	1	1	4
9	1	1	5
10	1	1	3
11	1	1	4
12	1	1	5
13	1	1	3
14	1	1	4
15	1	1	5

FIGURE 8.3

analyses on your own. We are going to be using a few different parts of the SPSS program for this analysis. However, running a factorial ANOVA in SPSS can be quite simple if you follow these steps:

1. *Click* **Analyze**.
2. *Select* **General Linear Model** (not **Generalized Linear Model**).
3. *Click* **Univariate**.
4. *Once you have clicked* **Univariate** *you will get the screen seen in Figure 8.4:*
5. *Highlight the dependent variable (in this example it is the conducting outcome) and move it into the* **Dependent Variable** *slot.*
6. *Highlight the two independent variables (in this example gender as well as highest music education degree) and move them into the* **Fixed Factor(s)** *slot. Remember that these are variables that are beyond your control and that do not change, so* **Fixed Factor(s)** *is where they belong.*
7. *Click on* **Options** *and make sure that you have highlighted each of the following:*
   - **Descriptive Statistics**
   - **Estimates of Effect Size**
   - **Homogeneity Tests**
8. *Click* **Continue**.
9. *Click* **OK** *to run the factorial ANOVA.*

Once you have clicked **OK**, you should get four boxes of information. As seen in Figure 8.5, the first box contains a listing of the between-subject variables and the number of participants in each group.

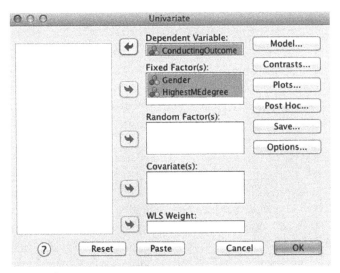

FIGURE 8.4

Between-Subject Factors

		N
Gender	0	79
	1	86
Highest ME degree	1	29
	2	82
	3	41
	4	13

FIGURE 8.5

Descriptive Statistics

Dependent Variable : Conducting Outcome

Gender	Highest ME degree	Mean	Std. Deviation	N
0	1	3.76	1.091	17
	2	6.02	.783	45
	3	8.07	.594	15
	4	8.50	.707	2
	Total	5.99	1.652	79
1	1	3.83	.835	12
	2	5.70	.777	37
	3	8.12	.766	26
	4	8.27	1.009	11
	Total	6.50	1.767	86
Total	1	3.79	.978	29
	2	5.88	.792	82
	3	8.10	.700	41
	4	8.31	.947	13
	Total	6.25	1.727	165

FIGURE 8.6

From this box we can see that 79 men and 86 women participated in the study. Of those participating, 29 had no music education degree, 82 had bachelor's degrees in music education, 41 held master's degrees in music education, and only 13 had completed the doctorate in music education.

The second box offers the descriptive statistics for each of these groupings (see Figure 8.6).

From this box, we can see how the two independent variables relate to each other through a cross-tabulation. If you want simple descriptive statistics for each individual group, you will need to run the descriptive statistics as discussed in previous chapters. As we look at Figure 8.5, we can see, for example, that the 17 men with no music education degree scored lower on the conducting outcome (3.76) than the 12 women with no music education degree (3.83).

The third box created in SPSS when you run a factorial ANOVA should look familiar to you. It is the Levene test for equality of error variance (see

Levene's Test of Equality of Error Variances[a]

Dependent Variable : Conducting Outcome

F	df1	df2	Sig.
1.524	7	157	.163

Test the null hypothesis that the error variance of the dependents variable is equal across groups.
[a]Design: Intercept + Gender + Highest ME degree + Gender* Highest ME degree

FIGURE 8.7

Figure 8.7). Although the title of the test might be slightly different, the outcomes are the same. As shown in Figure 8.7, the Levene test tests the null hypothesis that the error variance of the dependent variable is equal across all groups. That is, it tests to see if the variance of the conducting-outcome score is the same regardless of gender or degree. This is not to say it tests for similarities between the means. That is the job of the ANOVA. This test just tells us if the groups' variances are equal enough to meet the assumption of the homogeneity of variance for multiple groups on a single dependent outcome.

As we can see in Figure 8.7, the significance of the Levene test has a significance greater than .05. Therefore, we can proceed while we assume that there are equal variances and that we have met the required assumption for the ANOVA. Also, this assumption will inform us as to what type of post hoc test would be most appropriate to run if needed. As we have equal variance, we can run a post hoc test such as Scheffe or Tukey that also assumes equal variance. If not, we would have to run a post hoc test such as Dunnet or Games-Howell that does not assume equal variance.

The final box provided by SPSS gives us the results of the actual factorial ANOVA (see Figure 8.8).

From Figure 8.8, we can see that the information provided in the factorial ANOVA table is similar to that offered in the one-way ANOVA table (see Chapter 7, Figure 7.6). We can see that the overall model is significant (corrected model). Therefore, we know that something is going to be significant. We can also see that gender did not have a significant main effect on the conducting outcome. Highest earned music education degree, however, did have a significant impact on the conducting outcome. It is somewhat difficult to see in this example, but the column with gender and an asterisk, which indicates statistical significance, is the interaction row between gender and highest earned degree, $p = .587$. So of the three possible null hypotheses (one for the main effect of each independent variable, and one for the possible interaction) we will reject only one: the main effect of highest earned music education degree. However, even though we know we are going to reject one of the null

Tests of Between-Subjects Effects

Dependent Variable : Conducting Outcome

Source	Type III Sum of Squares	df	Mean Square	F	Sig.	Partial Eta Squared
Corrected Model	383.607[a]	7	54.801	81.396	.000	.784
Intercept	3079.591	1	3079.591	4574.140	.000	.967
Gender	.208	1	.208	.309	.579	.002
Highest ME degree	339.442	3	113.147	168.058	.000	.763
Gender* Highest ME degree	1.305	3	.435	.646	.587	.012
Error	105.702	157	.673			
Total	6944.000	165				
Corrected Total	483.309	164				

[a] R Squared = .784 (Adjusted R Squared = .774)

FIGURE 8.8

hypotheses, we do not have enough information to be able to tell where the actual differences exist between different levels of music education degrees.

Do we have enough information to know where the difference is? Not quite yet. As you can see from Figure 8.7, we did find a significant difference between different levels of music education degrees. However, we have no idea if the difference is between those with no degrees and doctoral degrees or between those with bachelor's degrees and master's degrees (or any other possible combination). In order to examine where actual difference exist, we need to conduct some post hoc tests.

The first step in conducting these tests is to decide of you have equal variance in your groups or not. As discussed above, as our test of homogeneity of error variance (Levene) found a significance level greater than .05, we can assume equal variance. This finding is important, as it will inform what type of post hoc test we should run.

To run a post hoc test in SPSS, follow these steps:

1. Click **Analyze**.
2. Select **General Linear Model** (not **Generalized Linear Model**).
3. Click **Univariate**.
4. Highlight the dependent variable (in this example it is the conducting outcome) and move it into the **Dependent Variable** slot.
5. Highlight the two dependent variables (in this example gender as well as highest music education degree) and move them into the **Fixed Factor(s)** slot. Remember that these are variables that are beyond your control and that do not change, so **Fixed Factor(s)** is where they belong.
6. Click **Options** and make sure that you have highlighted each of the following:

- **Descriptive Statistics**
- **Estimates of Effect Size**
- **Homogeneity Tests**

N.B. Steps 1 through 6 should be the exact same as your original analysis steps.

7. *Click* **Continue**.
8. *Click* **Post Hoc**.

You will see the screen that is shown in Figure 8.9.

10. *Select the variables (factors) that have more than two levels and that were found to be significantly different in the original analysis (in our example it is highest music education degree) and move them into the* **Post Hoc Tests for** *slot on the right.*
11. *You will then be able to select which post hoc test you want. In our example it is* **Scheffe**, *as we have met the assumption of homogeneity of variance.*
12. *Click* **Continue**.
13. *Click* **OK** *to run the analysis.*

You will get the same four boxes from the original analysis. You will, however, get an additional box with the post hoc test (see Figure 8.10).

FIGURE 8.9

Multiple Comparisons

Dependent Variable: Conducting Outcome
Scheffe

(I) Highest ME degree	(J) Highest ME degree	Mean Difference (I–J)	Std. Error	Sig.	95% Confidence Interval	
					Lower Bound	Upper Bound
1	2	−2.08*	.177	.000	−2.59	−1.58
	3	−4.30*	.199	.000	−4.87	−3.74
	4	−4.51*	.274	.000	−5.29	−3.74
2	1	2.08*	.177	.000	1.58	2.59
	3	−2.22*	.157	.000	−2.66	−1.78
	4	−2.43*	.245	.000	−3.12	−1.74
3	1	4.30*	.199	.000	3.74	4.87
	2	2.22*	.157	.000	−1.78	−2.66
	4	−.21	.261	.885	−.95	.53
4	1	4.51*	.274	.000	3.74	5.29
	2	2.43*	.245	.000	1.74	3.12
	3	.21	.261	.885	−.53	.95

Based on observed means.
The error term is Mean Square(Error)= .673.
*The mean difference is significant at the

FIGURE 8.10

From Figure 8.10, we can see that significant differences existed between each of the groups except for between-groups 3 and 4 (those with master's degrees and those with doctoral degrees in music education). In order to tell which groups were higher than which groups, we would have to return to the descriptive-statistics box offered in the original factorial ANOVA.

In Example 8.4, you can see an example of a write-up of a post hoc test.

You will note that in Example 8.4, the author did not write out the descriptive statistics. Instead, the researcher reported the descriptive information in a table.

Is this enough information to begin to write the report of the findings? Yes. From the analysis thus far, you can report all of the descriptive statistics, the pertinent information regarding each of the main effects and interactions, and the effect size (eta squared) that was calculated for you in SPSS. You can also report the post hoc test findings and clarify them by using the descriptive statistics (mean and standard deviation).

Reporting a Factorial ANOVA

Once you have conducted all of the necessary statistical procedures and post hoc tests, you now have enough information to write a research report that can give the reader enough information to evaluate and interpret your work. Given the grouping of the participants in our running example, one possible write-up of the data may look like Example 8.5.

EXAMPLE 8.4 POST HOC WRITE-UP EXAMPLE

Scheffé post hoc comparisons evidenced significant mean score differences on each individual subtest between students participating in band and those participating in choir (all $p < .01$) and between students participating in band and those not participating in music (all $p < .01$). In all instances, students participating in band scored significantly higher than those in choir or not participating in music (see Figure 1). No significant differences were found between choir and nonparticipant mean scores for any subtest on either the fourth- or sixth-grade proficiency tests.

Kinney, D. W. (2008). Selected demographic variables, school music participation, and achievement test scores of urban middle school students. *Journal of Research in Music Education, 56*(2), 145–161.

EXAMPLE 8.5 POSSIBLE FACTORIAL ANOVA WRITE-UP

In order to examine the main effects of gender and highest music education degree earned, as well as any possible interaction on participants' ($N = 165$) conducting outcome score, I computed a 2 × 4 factorial ANOVA. Prior to the analysis, I established the homogeneity of variance by using the Levene test ($F = 1.52, p = .16$). I used gender and highest earned degree at categorical independent variables and participants' scores on the conducting outcome as the dependent variable and found no significant interaction effect ($F = .646, df = 3, p = .587$). I found no significant main effect for gender ($F = .309, df = 1, p = .579$), but did find a significant main effect for highest earned music education degree ($F = 168.06, df = 3, p < .001, \eta^2 = .763$). The high effect size, as indicated by eta squared, suggests practical as well as statistical significance. I conducted a Scheffe post hoc test to determine where the significant differences existed. Each group scored significantly higher as the participants' degrees increased. Those with bachelor's degrees ($n = 82, M = 5.88, SD = .79$) scored higher than those without music education degrees ($n = 29, M = 3.79, SD = .98$). Participants with master's degree in music education ($n = 41, M = 8.10, SD = .70$) scored significantly higher than those with bachelor's degrees or without music education degrees. Similarly, those with doctoral degrees ($n = 13, M = 8.31, SD = .95$) scored significantly higher than those without degrees or with bachelor's degrees. No significant difference existed, however, between those with master's degrees and those with doctoral degrees.

Writing Hints for Factorial ANOVA

From the research report in Example 8.5, we see some hints that might help you state your findings as clearly as possible while providing all of the necessary information.

- Make sure that you provide the following in your report:
  - Levene finding
  - N
  - n
  - F for each test (including Levene)
  - df for each test
  - p for each comparison
  - η^2 for each significant finding
  - Discussion of practical and statistical significance
  - Means and standard deviations of all groups discussed
  - What type of factorial ANOVA (i.e., 2 × 4, etc.)
  - Post hoc tests you used
- Do not repeat the means of groups more than once. One you have reported it, you do not need to do so again.
- Another way to report the means and standard deviations of the groups could be in a table.
- Per American Psychological Association guidelines, use first-person pronouns for clarity, especially when describing actions you took, while keeping all of your actions in past tense.
- Use the active voice.
- Avoid personification.
- Make sure all statistical indicators are in italics.
- Keep your actions as researcher in the past tense.

Factorial ANOVA Exercises

Given the second provided dataset, Factorial ANOVA Example 2 ▶, compute the most logical analysis. In this example, the columns are

- Treatment (0 = Control Group, 1 = Treatment Group) The treatment is pre-rehearsal group stretching exercises.
- Instrument/Voice (0 = Vocalist, 1 = Instrumentalist)
- Score on a self-reported physical discomfort scale (0 = no discomfort, 5 = great discomfort)

In this dataset, a researcher is interested in the effect of stretching on physical discomfort between both instrumentalists and vocalists. The two independent variables are treatment (those who stretched and those who did not) and instrumentalist and vocalist. The dependent variable is the score on scale of discomfort. Therefore, this test would be a 2 × 2 factorial ANOVA.

Once you have computed the most appropriate statistical test, write a report that includes any descriptive statistics necessary to understand the factorial ANOVA, as well as the test itself and, finally, effect size.

9 | Multivariate Analysis of Variance (MANOVA)

Use of the Multivariate Analysis of Variance (MANOVA)

What is a researcher to do when he or she wants to create a statistical model in order to compare the means of more than one dependent variable for different groups? As we have seen so far, t-tests must have only two groups or levels. A one-way ANOVA can have multiple groups or levels for the independent variable, but still only one independent variable and one dependent variable. The factorial ANOVA can handle more than one independent variable, but still has room in the model for only one dependent variable. What do all of these tests from previous chapters have in common? They are univariate tests. That is, they can include only one *dependent* variable. So in order to answer the first question about what a researcher can do to examine more than one dependent variable, we must look toward multivariate analysis. It happens that the M in MANOVA stands for multivariate. A researcher can use a MANOVA to explore how any number of categorical independent variables impact two or more dependent variables.

Initially, you might ask, "Why not run multiple ANOVAs rather than one MANOVA?" There are two primary answers to this question. First, running multiple analyses increases the threat of a Type I error. The more parsimonious statistical design (the one that does what you need done with the fewest number of comparisons) is always preferable. However, keep in mind that using more advanced statistical analyses such as a MANOVA does not take the place of a well-designed study or adequate data collection.

As seen in Table 1.6 in Chapter 1, researchers use a MANOVA to answer a single research question: do differences exist between two or more groups on two or more dependent variables?

For example, a music education researcher may be interested in the differences, based in which type of traditional ensemble class the students participated, between high school music students' enjoyment of music class and their audition scores. Does a difference exist among band, orchestra, and choir students with regard to their enjoyment of music class and their audition scores? In this example, the categorical independent variable is the class in which students enrolled. The two dependent variables are their self-reported levels of enjoyment and their audition scores.

Although we do not discuss it at length in this chapter or book, a MANOVA can be factorial as well. That is, a researcher can have multiple independent variables and dependent variables. In such cases, the researcher is interested in both the main effects and the interactions, as discussed in Chapter 8. For example, in Example 9.1, the researcher employed a MANOVA with three dependent variables and three independent variables and found only one main effect. In this example, the three categorical independent variables are gender (two levels), school level (two levels), and instrument (five levels). Therefore, this example is a 2 × 3 × 5 factorial MANOVA. After finding the single main effect of gender on each of the dependent variables, the researcher conducted subsequent ANOVAs in order to find if gender did in fact significantly impact all three dependent variables.

EXAMPLE 9.1 REPORTING OF MANOVA

The author analyzed data in a MANOVA-type design by using the general linear model in SPSS to test the four research hypotheses on the dependent-variable confidence, anxiety, and attitude. Independent variables identified were gender—two levels: male ($n = 83$) and female ($n = 54$); school level—three levels: middle school ($n = 50$), high school ($n = 43$), and college/adult ($n = 44$); and instrument—five levels: saxophone ($n = 33$), trombone ($n = 22$), trumpet ($n = 31$), rhythm section ($n = 24$), and other ($n = 19$)]. Univariate subanalysis of variance tests followed to identify dependent variables contributing to any effect.

The results of the MANOVA design indicated a single main effect for gender, rejecting the null hypothesis of multivariate equality of means over all groups, $F(3, 105) = 3.94$, $p < .01$. Follow-up univariate tests indicate that the effect of gender was attributable to all three dependent variables: confidence, $F(1, 107) = 9.67$, $p < .01$; anxiety, $F(1, 107) = 10.09$, $p < .01$; and attitude, $F(1, 107) = 4.79$, $p < .05$. Therefore, the author rejected the null hypotheses that males and females would not differ in confidence, anxiety, and attitude toward learning jazz improvisation. Means and standard deviations for the three dependent variables are summarized in Tables 1, 2, and 3.

Wehr-Flowers, E. (2006). Differences between male and female students' confidence, anxiety, and attitude toward learning jazz improvisation. *Journal of Research in Music Education, 54*(4), 337–349.

In Example 9.1, the researcher focused solely on the *F* statistic for each individual analysis within the MANOVA. Some researchers, in order to indicate that the overall model was significant, elect to report the lambda (λ) and its corresponding *F* statistic. With a MANOVA, the lambda is often referred to as the *omnibus test*, which looks for significance in the overall statistical model. Lambda, in a MANOVA, tells us whether or not the overall model is significant and worth examining further. It does not explain where or to what extent any differences exist. What we are most interested in is a statistic called Wilks's lambda, and the *F* value associated with that. Lambda is a measure of the percent of variance in the dependent variables that is not explained by differences in the independent variable. Lambda values range between 0 and 1. The closer to zero means that there is no variance that is not explained by the independent variable. Lambda (most often Wilks's lambda) should be reported so that the reader has a better understanding of the overall practical significance found in the model.

In Example 9.2, researchers report the lambda as well as the findings of the subsequent analyses. Once the researchers had found that the overall test was significant (the Wilks's λ), they conducted subsequent ANOVAs on each of the independent variables to find where significance existed. Because neither independent variable had more than two levels, no post hoc tests were needed. Finally, the authors report the effect size and the subsequent argument for practical significance.

If, however, the independent variables found to be significantly related to any dependent variable have more than two levels, researchers need to conduct post hoc tests to see exactly where the differences exist. Running these post hoc tests for a MANOVA is not different from running the ANOVAs discussed in previous chapters. See Example 9.3 for an example of reporting post hoc tests for a MANOVA.

Assumptions of MANOVA

The MANOVA has many of the same assumptions as all of the different form of ANOVAs discussed in previous chapters. It is different, however, in that the MANOVA can handle more than one dependent variable. The assumptions of the MANOVA are the following:

1. The data collected for the dependent variables are continuous rather than categorical. Remember that we need to have at least two dependent variables.
2. The data collected for the independent variable is categorical rather than continuous.
3. Each observation is independent of any other observation.
4. Multivariate normality—all of the dependent variables are normally distributed themselves and that any combination of the dependent variables is normally distributed

EXAMPLE 9.2 REPORTING OF LAMBDA IN A MANOVA

Multivariate analysis of variance (MANOVA) was used to determine whether differences in grade weights might be attributed to school context variables such as instructional level (middle school/junior high, high school) and teaching specialization (instrumental, choral). For the multivariate analysis of school context effects, teaching level (middle school, high school) and teaching specialization (instrumental, choral) served as the grouping variables, while the dependent variable set included grade weights corresponding to attendance, attitude, practice, knowledge, and skill (see Table 5).

The MANOVA revealed a significant interaction effect for teaching level and teaching specialization ($\Lambda = 0.93$, $p < .001$), as well as significant main effects for teaching level ($\Lambda = 0.90$, $p < .001$) and teaching specialization ($\Lambda = 0.74$, $p < .001$). Follow-up univariate tests determined which mean differences in grading criteria weights contributed to the significant multivariate outcome.

Significant interactions existed for two dependent variables—knowledge ($F = 7.73$, $p < .01$), and practice ($F = 16.72$, $p < .001$). Simple main effects tests were used to clarify the nature of the interactions. Middle school choral directors gave significantly more weight ($M = 16.9\%$) to written assessment of musical knowledge than did middle school instrumental directors ($M = 9.6\%$), while there was no significant difference in the amount of weight given to musical knowledge by high school instrumental and choir directors. Conversely, middle school band directors gave significantly more weight to practice assessments ($M = 11.9\%$) than middle school choir directors ($M = 2.6\%$). As with knowledge assessment, however, there was no significant difference in the amount of weight given to practice by high school instrumental and choir directors.

Significant main effects, based on teaching level, emerged for attendance. High school directors gave greater weight to attendance ($M = 28.2\%$) than middle school directors ($M = 18.5\%$). Significant main effects, based on teaching specialization, emerged for attitude and performance. Choral directors gave greater weight to attitude ($M = 37.5\%$) than instrumental directors ($M = 21.0\%$), while instrumental directors gave greater weight to performance assessments of musical skill ($M = 31.5\%$) than choral directors ($M = 21.2\%$). While numerous significant differences in mean weights assigned to various assessment/grading criteria emerged, it is important to note that all effect sizes (η^2) for the univariate analyses were less than .10, a finding that suggests that these particular contextual variables (reflecting primary teaching positions or responsibilities) are accounting for only a small portion of the variance in grade weights.

Russell, J. A., & Austin, J. R. (2010). Assessment practices of secondary music teachers. *Journal of Research in Music Education, 58*(1), 37–54.

> **EXAMPLE 9.3 REPORTING OF POST HOC TESTS IN MANOVA**
>
> The MANOVA indicated that the tone quality of the stimuli was a significant factor: $F(4, 53) = 92.86, p < .001$. Tone-quality and pitch-rating means and standard deviations for each of the tone-quality conditions are presented in Table 2. Subsequent univariate analysis indicated that ratings of the three tone-quality conditions were significantly different from one another—$F(2, 112 = 311.08, p < .001)$. Scheffe analysis indicated significant differences between each of the tone-quality rating means ($p < .001$). Overall pitch rating means were also significantly different—$F(2, 112) = 60.54, p < .001$). Scheffe analysis also indicated significant differences between all pitch-rating means ($p < .001$).Table 2 shows that bright tone quality was rated significantly brighter in tone quality and sharper in pitch, whereas dark tone quality was rated significantly darker in tone quality and flatter in pitch.
>
> ---
>
> Worthy, M. D. (2000). Effects of tone-quality conditions on perception and performance of pitch among selected wind instrumentalists. *Journal of Research in Music Education, 48*(3), 222–236.

5. Homogeneity of the covariance matrices—each of the dependent variables has equal variance when compared to each independent variable. This assumption is similar to the assumption of homogeneity of variance for a univariate test as assessed by the use of a Levene test. In a MANOVA, however, we use a test called Box M (more on that later).
6. Linearity—there is a linear relationship between independent variables and dependent variables.

Meeting the Assumptions of MANOVA

Although the MANOVA is robust against a few of the assumptions (mainly the issue of multivariate normality), it is important to ensure that we use appropriate statistical tests with data that meet the assumptions so that our conclusions can be based upon defensible analyses. Making sure to use only categorical independent variables and continuous dependent variables is an easy first step to ensuring a good MANOVA analysis, as is collecting independent observations. Although it is rather difficult to explicitly offer data for this assumption, you can check the normality by using kurtosis and skewness indices (remember that no skew and perfect kurtosis indices are 0). The homogeneity of the covariance-matrices assumption, however, requires a Box M test.

The Box M test works similarly to the Levene test in a univariate analysis. Researchers use the Box M test to test whether or not the covariance matrices of the dependent variables are significantly different across levels of the independent variable. If Box's M is significant, it means you have violated an assumption of MANOVA (much as a significant finding of a Levene test is not a good thing). There is good news, however. A significant Box M test does not mean that you must necessarily abandon the analysis. The Box M is a very conservative test and the MANOVA is often considered robust against a significant finding if you have equal cell sizes (roughly the same number of participants in each grouping in the analysis) and large N. If you do not have equal cell sizes or a large N, the violation of this assumption means that you should find an alternative analysis or collect more data.

Research Design and the MANOVA

The MANOVA can be used in any research design in which the researcher has collected data on more than one dependent variable as well as one or more independent variables. Keep in mind that although the dependent variables should be somewhat correlated, they should not be highly correlated. If they are highly correlated, there is relatively little additional information you will gain from the analysis. The MANOVA builds upon the ANOVA as it does much of the same, but can be used to create a slightly more complex analysis by examining the impact of one or more than one categorical independent variable on more than one dependent variable. One way to visualize a MANOVA design is

$O_{\text{dependent variable 1}}$ $O_{\text{dependent variable 2}}$

$O_{\text{dependent variable 1}}$ $O_{\text{dependent variable 2}}$

$O_{\text{dependent variable 1}}$ $O_{\text{dependent variable 2}}$

In this design, three groups are being observed for two different dependent variables. The grouping is the categorical independent variable. In our musical example of choir, band, and orchestra students, the design may look like

Choir $O_{\text{class enjoyment}}$ $O_{\text{audition score}}$

Band $O_{\text{class enjoyment}}$ $O_{\text{audition score}}$

Orchestra $O_{\text{class enjoyment}}$ $O_{\text{audition score}}$

In this example, the grouping as to what class students enrolled in is the independent variable, and the self-reported class enjoyment and audition scores are the dependent variables. This is not to say that MANOVA cannot be used in an experimental design. A similar MANOVA in an experimental design may be visualized as

$O_{\text{predependent variable 1}}$ $O_{\text{predependent variable 2}}$	X_1	$O_{\text{postdependent variable 1}}$ $O_{\text{postdependent variable 2}}$
$O_{\text{predependent variable 1}}$ $O_{\text{predependent variable 2}}$	X_2	$O_{\text{postdependent variable 1}}$ $O_{\text{postdependent variable 2}}$
$O_{\text{predependent variable 1}}$ $O_{\text{predependent variable 2}}$	X_3	$O_{\text{postdependent variable 1}}$ $O_{\text{postdependent variable 2}}$

In this example, the grouping variable is still the independent variable. However, this time, each group received some type of treatment. The researcher also collected data on two dependent variables prior to the treatment and following the treatment. A researcher could analyze this data in a few ways. The two most common would be to compute the mean gain score for each dependent variable and use that as the two scores to analyze. The other would be to use the scores of the pretests as a covariate. If a researcher does this, the analysis is actually a MANCOVA, which will be discussed later in this text.

For a musical example of this possible design, we might consider our initial example, but rather than break students apart by their class, we might give students three different types of rehearsal experiences:

1. A traditional, teacher-led rehearsal.
2. A more student-centered rehearsal experience with greater student input.
3. A student-led rehearsal in which students make all musical decisions.

In this case, the design could be visualized as

$O_{\text{preenjoyment}}$ $O_{\text{preaudition scores}}$	$X_{\text{traditional rehearsal}}$	$O_{\text{postenjoyment}}$ $O_{\text{postaudition scores}}$
$O_{\text{preenjoyment}}$ $O_{\text{preaudition scores}}$	$X_{\text{student-centered rehearsal}}$	$O_{\text{postenjoyment}}$ $O_{\text{postaudition scores}}$
$O_{\text{preenjoyment}}$ $O_{\text{preaudition scores}}$	$X_{\text{student led rehearsal}}$	$O_{\text{postenjoyment}}$ $O_{\text{postaudition scores}}$

The Null Hypotheses for MANOVA

Researchers can use a MANOVA to test the hypothesis that one or more independent variables, or factors, have an effect on two or more dependent variables. These null hypotheses could be stated as

$H_0 : \mu_1 = \mu_2 = \mu_3$ etc. for Dependent Variable 1
$H_0 : \mu_1 = \mu_2 = \mu_3$ etc. for Dependent Variable 2
Etc.

For our musical example, the null hypotheses could be stated as

$H_0 : \mu_{choir} = \mu_{band} = \mu_{orchestra}$ for class enjoyment
$H_0 : \mu_{choir} = \mu_{band} = \mu_{orchestra}$ for audition scores

The alternative hypotheses for the MANOVA would be

$H_a : \mu_1 \neq \mu_2 \neq \mu_3$ etc. for Dependent Variable 1
$H_a : \mu_1 \neq \mu_2 \neq \mu_3$ etc. for Dependent Variable 2
Etc.

For our musical example, the alternative hypothesis would be

$H_a : \mu_{choir} \neq \mu_{band} \neq \mu_{orchestra}$ for class enjoyment
$H_a : \mu_{choir} \neq \mu_{band} \neq \mu_{orchestra}$ for audition scores

The directional hypothesis for the MANOVA could be stated as

$H_1 : \mu_1 < \mu_2 < \mu_3$ etc. for Dependent Variable 1
$H_1 : \mu_1 < \mu_2 < \mu_3$ etc. for Dependent Variable 2

In our musical example, the directional hypothesis might be (depending on information from previous research)

$H_a : \mu_{choir} < \mu_{band} < \mu_{orchestra}$ for class enjoyment
$H_a : \mu_{choir} < \mu_{band} < \mu_{orchestra}$ for audition scores

Setting Up a Database for a MANOVA

The database for a MANOVA should have at least three columns. At least two columns of continuous dependent variables are needed with at least one column of a categorical independent variable. The independent-variable column should look just like an independent-variable column you created for an ANOVA or factorial ANOVA. That is, it should be a grouping variable. In Figure 9.1 we see a database created to run a simple MANOVA with two dependent variables and one independent variable. The independent variable is the music class that high school students have enrolled in (1 = Choir, 2 = Band, 3 = Orchestra). The two dependent variables are their ratings of enjoyment (1 = does not enjoy class, 5 = really enjoys music class) and an audition score (possible range of 0–100).

Conducting a MANOVA in SPSS

For this section, please use the provided example database, MANOVA Example 1 ⓘ, in order to complete the analysis yourself while reading. This exercise will help you better understand through experience. Conducting a

	musicclass	enjoyment	audition
1	1	2	88
2	1	3	87
3	1	2	87
4	1	3	86
5	1	3	86
6	1	3	85
7	1	3	84
8	1	2	83
9	1	2	84
10	1	2	85
11	1	3	86
12	1	3	87
13	1	2	86
14	1	3	85
15	1	2	84

FIGURE 9.1

MANOVA in SPSS will look rather similar to running a factorial ANOVA and can be accomplished by following these steps:

1. Click **Analyze**.
2. Click **General Linear Model** (not **Generalized Linear Model**).
3. Click **Multivariate** *(remember that we have more than one dependent variable)*.
4. Once you have clicked **Multivariate** *you will get the screen shown in Figure 9.2:*
5. *Highlight the dependent variables (in this example it is the* **enjoyment** *score and* **audition** *score) and move it into the* **Dependent Variables** *slot.*
6. *Highlight the independent variable (in this example* **musicclass***) and move it into the* **Fixed Factor(s)** *slot.*
7. *Click* **Options** *and make sure that you have selected each of the following:*
   - **Descriptive statistics**
   - **Estimates of effect size**
   - **Homogeneity tests**
8. *Click* **Continue**.
9. *Click* **OK** *to run the MANOVA.*

Once you have clicked **OK**, you should get six boxes of information. As seen in Figure 9.3, the first box simply contains a listing of the between-subject variables and the number of participants in each group.

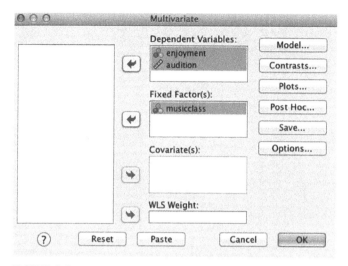

FIGURE 9.2

Between-Subjects Factors

		N
music class	1	33
	2	33
	3	34

FIGURE 9.3

From this box, we can see that 33 students were enrolled in choir (group 1) as well as band (group 2), while 34 students were enrolled in orchestra (group 3).

The second box that SPSS creates shows the descriptive statistics for each group of students as it compares to each dependent variable (see Figure 9.4).

From this box we can see the means and standard deviations of each group in regard to enjoyment as well as the group's audition scores. We can see, for instance, that the band students expressed the highest mean of enjoyment and that the lowest audition scores came from the choir students.

The third box that SPSS creates gives the researcher the Box M test results (see Figure 9.5). Remember that this test is analogous to the Levene test and tells the researcher if she has met the assumption of homogeneity of covariance matrices.

In Figure 9.5, we see that the Box M test is not significant ($p > .05$), meaning that we have met the assumption of equality or homogeneity of covariance matrices and can continue with the analysis.

The next provided box (see Figure 9.6) gives us the actual outcome of the MANOVA. In this table, the researcher is provided with the Wilks's lambda and its corresponding F and significance as well as the partial eta squared.

Descriptive Statistics

	music class	Mean	std. Deviation	N
enjoyment	1	2.52	.508	33
	2	4.48	.508	33
	3	3.56	.504	34
	Total	3.52	.948	100
audition	1	85.52	1.734	33
	2	96.52	1.121	33
	3	96.50	1.108	34
	Total	92.88	5.364	100

FIGURE 9.4

Box's Test of Equality of Covariance Matrices[a]

Box's M	9.223
F	1.491
df1	6
df2	233257.676
Sig.	.177

[a] Tests the null hypothesis that the observed covariance matrices of the dependent variables are equal across groups.

FIGURE 9.5

Multivariate Tests[a]

Effect		Value	F	Hypothesis df	Error df	sig.	Partial Eta squared
Intercept	Pillai's Trace	1.000	241705.26[b]	2.000	96.000	.000	1.000
	Wilks' Lambda	.000	241705.26[b]	2.000	96.000	.000	1.000
	Hotelling's Trace	5035.526	241705.26[b]	2.000	96.000	.000	1.000
	Roy's Largest Root	5035.526	241705.26[b]	2.000	96.000	.000	1.000
music class	Pillai's Trace	1.272	84.739	4.000	194.000	.000	.636
	Wilks' Lambda	.035	208.028[b]	4.000	192.000	.000	.813
	Hotelling's Trace	18.713	444.424	4.000	190.000	.000	.903
	Roy's Largest Root	18.233	884.317[c]	2.000	97.000	.000	.948

FIGURE 9.6

We can see in Figure 9.6 that SPSS offers four different means of evaluating the significance of the overall MANOVA model: Pillai's trace, Wilks's lambda, Hotelling's trace, and Roy's largest root. Although we will focus on the use of Wilks's lambda, it is not the only or always most appropriate test to use. Some general guidelines follow:

- Wilks's lambda offers an exact F statistic and is most common (therefore most readily understood), as it is equal to the variance not accounted for by the combined dependent variables.
- Pillai's trace considers pooled effect variance and offers the most conservative F statistic (which can be seen in Figure 9.6). This test is often considered the most robust.
- Hotelling's trace shows the pooled ratio of effect variance to error variance and is often similar to Roy's largest root.
- Roy's largest root is based upon the largest eigenvalue found and creates an upper-bound F statistic. This index should be ignored if the other tests are not significant.

In order to interpret the table in Figure 9.6, we focus on the row that indicates the independent variable (in this case music class). We see that the Wilks's lambda equals .035 with an F of 208.03, significance of $< .001$, and a partial eta squared of .81. The significant lambda tells us that the overall model is significant, meaning that somewhere we have found some difference based upon the music class variable. We must look to the test of between-subject outcomes in the next two boxes to know where differences exist.

The fifth box created by SPSS will look very familiar to you. It is a Levene test of equality of error variance (see Figure 9.7). Remember that Box M told us that we met the assumption of the MANOVA or the omnibus test; this Levene test is there to tell us if we met the assumption of the follow-up ANOVAs automatically conducted within the MANOVA.

From this box we can examine the equality of error variance of both dependent variables. We see that we have met the assumption of equal variance for the enjoyment dependent variable, but have violated the assumption of equal variance for the audition dependent variable. This lack of assumption will be something to consider if we need to conduct any post hoc tests based upon the ANOVAs to follow.

The final box that SPSS creates in this analysis is the table of ANOVAs of each individual analysis (see Figure 9.8).

Again, we are most interested in the row that is labeled with our independent variable (musicclass). From this row, we can see that both enjoyment

Levene's Tesi of Equality of Error Variances[a]

	F	df1	df2	Sig.
enjoyment	.346	2	97	.708
audition	3.956	2	97	.022

Tests the null hypothesis that the error variance of the dependent variable is equal across groups.
[a]Design: Intercept + musicclass

FIGURE 9.7

Tests of Between-Subjects Effects

Source	Dependent Variable	Type III Sum of Squares	df	Mean Square	F	Sig.	Partial Eta Squared
Corrected Model	enjoyment	64.093a	2	32.046	125.004	.000	.720
	audition	2671.575b	2	1335.788	732.104	.000	.938
Intercept	enjoyment	1238.519	1	1238.519	4831.115	.000	.980
	audition	861819.639	1	861819.639	472337.071	.000	1.000
music class	enjoyment	64.093	2	32.046	125.004	.000	.720
	audition	2671.575	2	1335.788	732.104	.000	.938
Error	enjoyment	24.867	97	.256			
	audition	176.985	97	1.825			
Total	enjoyment	1328.000	100				
	audition	865518.000	100				
Corrected Total	enjoyment	88.960	99				
	audition	2848.560	99				

a R Squared = .720 (Adjusted R Squared = .715)
b R Squared = .938 (Adjusted R Squared = .937)

FIGURE 9.8

and audition scores were impacted in some way by what music class in which students were enrolled. We see that, for instance, the *F* statistic for enjoyment is 125.00 with a significance of <.001 and a partial eta squared of .72. This figure should look just like an ANOVA table to you, because it is one.

Do we have enough information to know where the difference is? Not quite yet. As you can see from Figure 9.8, we found a significant difference between music classes on both dependent variables. However, we have no idea if the difference is between band kids and choir kids, orchestra kids and choir kids, or band kids and orchestra kids. In order to examine where actual difference exists, we need to conduct some post hoc tests. We are going to do this test almost exactly as we did for any ANOVA.

The first step in conducting these tests is to decide if you have equal variance in your groups or not. As discussed above, our test of homogeneity of error variance (Levene) found a significance level greater than .05 as well as one smaller. Therefore, we can assume equal variance for one post hoc test, but not the other.

To run a post hoc test in SPSS, follow these steps:

1. *Click* **Analyze**.
2. *Click* **General Linear Model** (not **Generalized Linear Model**).
3. *Click* **Multivariate**.
4. *Highlight the dependent variables (in this example it is the* **enjoyment** *score and* **audition** *score) and move it into the* **Dependent Variables** *slot.*
5. *Highlight the independent variable (in this example* **musicclass***) and move it into the* **Fixed Factor(s)** *slot.*

6. *Click on* **Options** *and make sure that you have highlighted each of the following:*
   - **Descriptive statistics**
   - **Estimates of effect Size**
   - **Homogeneity tests**
7. *Click* **Continue**.

N.B.: Steps 1 through 7 should be the exact same as your original analysis steps

8. *Click* **Continue**.
9. *Click* **Post Hoc**. *You will see the screen shown in Figure 9.9.*
10. *Select the variables (factors) that you would like to include in the post hoc analysis (in our example it is* **musicclass***) and move it into the* **Post Hoc Tests for** *slot.*
11. *You will then be able to select which post hoc test you want. In our example it is* **Scheffe** *(under* **Equal Variances Assumed***) and* **Games-Howell** *(under* **Equal Variances Not Assumed***), as we have met the assumption of homogeneity of variance for one test but not the other.*
12. *Click* **Continue**.
13. *Click* **OK** *to run the analysis.*

You will get the same six boxes from the original analysis. You will, however, get an additional box with the post hoc test (see Figure 9.10).

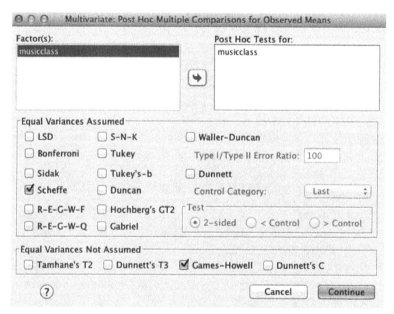

FIGURE 9.9

Multiple Comparisons

Dependent Variable		(I) music class	(J) music class	Mean Difference (I-J)	Std. Error	Sig.	95% Confidence Interval	
							Lower Bound	Upper Bound
enjoyment	Scheffe	1	2	-1.97*	.125	.000	-2.28	-1.66
			3	-1.04*	.124	.000	-1.35	-.74
		2	1	1.97*	.125	.000	1.66	2.28
			3	.93*	.124	.000	.62	1.23
		3	1	1.04*	.124	.000	.74	1.35
			2	-.93*	.124	.000	-1.23	-.62
	Games-Howell	1	2	-1.97*	.125	.000	-2.27	-1.67
			3	-1.04*	.124	.000	-1.34	-.75
		2	1	1.97*	.125	.000	1.67	2.27
			3	.93*	.124	.000	.63	1.22
		3	1	1.04*	.124	.000	.75	1.34
			2	-.93*	.124	.000	-1.22	-.63
audition	Scheffe	1	2	-11.00*	.333	.000	-11.83	-10.17
			3	-10.98*	.330	.000	-11.81	-10.16
		2	1	11.00*	.333	.000	10.17	11.83
			3	.02*	.330	.999	-.81	.84
		3	1	10.98*	.330	.000	10.16	11.81
			2	-.02*	.330	.999	-.84	.81
	Games-Howell	1	2	-11.00*	.360	.000	-11.87	-10.13
			3	-10.98*	.357	.000	-11.84	-10.13
		2	1	11.00*	.360	.000	10.13	11.87
			3	.02*	.272	.998	-.64	.67
		3	1	10.98*	.357	.000	10.13	11.84
			2	-.02*	.272	.998	-.67	.64

Based on observed means.
The error term is Mean Square(Error) = 1.825.

FIGURE 9.10

In these post hoc tests, we use the Scheffe analysis in the test that met the assumption of equal variance and the Games-Howell in the test that did not meet the assumption of equal variance. In our example, enjoyment did meet the assumption, while audition did not. Therefore, we use the Scheffe for enjoyment and Games-Howell for the audition analysis. From these post hoc tests, we can see that a significant difference existed between all groups in regard to enjoyment of their music class. In regard to audition scores, however, we see that all groups were significantly different except for groups 2 and 3 (orchestra and band students). In order to see who scored higher or lower, we simply return to the descriptive statistics offered in the original analysis and report means and standard deviations.

Is this enough information? Yes. From the analysis thus far, you can report all of the descriptive statistics, the significance of the overall model (lambda), the pertinent information regarding each of the main effects and interactions, and the effect size (partial eta squared) that was calculated for you in SPSS. You can also report the post hoc test findings and clarify them by using the descriptive statistics (mean and standard deviation).

Reporting a MANOVA

Once you have conducted the omnibus test and any required post hoc tests for this example, you should have all the information you need to complete a research report that will give the readers all of the information they need to understand, critique, and apply your findings appropriately. Given the dataset in our running example, one possible write-up of the data may look like Example 9.4.

Writing Hints for MANOVA

From the write-up in Example 9.4, we see some hints that might help you state your findings as clearly as possible while providing all of the necessary information.

- Make sure that you provide the following in your report:
    o Lambda and its corresponding F, significance, and effect size
    o Box M for the MANOVA assumption
    o Levene finding for each univariate follow-up test
    o N
    o n
    o F for each test (including Levene)
    o df for each test
    o p for each comparison
    o η^2 for each significant finding

- Discussion of practical and statistical significance
- Means and standard deviations of all groups discussed
- Post hoc tests you used and why
- Do not repeat the means of groups more than once. One you have reported it, you do not need to do so again.
- Another way to have reported the means and standard deviations of the groups could have been in a table.
- Per American Psychological Association guidelines, use first-person pronouns for clarity, especially when describing actions you took, while keeping all of your actions in past tense.
- Use the active voice.
- Avoid personification.
- Make sure all statistical indicators are in italics.
- Keep your actions as researcher in the past tense.

EXAMPLE 9.4 POSSIBLE MANOVA REPORT

In order to examine the impact of high school students' ($N = 100$) music class enrollment on their self-reported enjoyment of music class and audition scores, we conducted MANOVA by using music class enrollment as the independent variable (choir, band, orchestra). We established the equality of covariance matrices by using the Box M test (Box M = 9.22, $p = .18$) and found that the overall MANOVA was significant ($\Lambda = .035$, $F = 208.03$, $p < .001$). In order to determine which mean differences contributed to the significant multivariate outcome, we conducted follow-up univariate tests. We attempted to established the assumption of equal variance by using the Levene test and found that the enjoyment variable met the assumption ($F = .357$, $df = 2$, $p = .71$), while the audition variable did not ($F = 3.96$, $df = 2$, $p = .02$). We found both enjoyment ($F = 125.00$, $df = 2$, $p < .001$) and audition scores ($F = 732.10$, $df = 2$, $p < .001$) were both significantly related to participants' music class enrollment and explained a large amount of variance ($\eta^2 = .72$ and .94 respectively). Finally, we conducted a Scheffe post hoc test to determine where the significant differences existed in regard to enjoyment. Each group differed significantly. Band students ($n = 33$, $M = 4.48$, $SD = .51$) reported higher enjoyment than choir students ($n = 33$, $M = 2.52$, $SD = .51$) and orchestra students ($n = 34$, $M = 3.56$, $SD = .50$). Orchestra students enjoyed their music class more than choir students as well. Because of the significant Levene test finding for the audition score variable, we employed a Games-Howell post hoc test and found significant differences between each group except between band and orchestra students. Band students ($n = 33$, $M = 96.51$, $SD = 1.12$) scored significantly higher audition scores than choir students ($n = 33$, $M = 5.88$, $SD = .79$). Orchestra students ($n = 34$, $M = 96.50$, $SD = 1.11$) also scored significantly higher audition scores than choir students.

MANOVA Exercises

Given the second provided dataset, MANOVA Example 2 ▶, compute the most logical analysis. In this example, the columns are

- Dependent Variable 1: Music Aptitude Profile score
- Dependent Variable 2: Sight Reading Score
- Independent Variable: Geographic Region (1 = Northeast, 2 = Southeast, 3 = Midwest, 4 = Southwest, 5 = Northwest)

This analysis will have you thinking like a university professor thinking about the enrollment decisions you will have to make on the basis of often-collected data.

Once you have computed the most appropriate statistical tests, write a report that includes any descriptive statistics necessary to understand the MANOVA, as well as the test itself and, finally, effect size.

10 | Repeated-Measures Analysis of Variance

Use of Repeated-Measures ANOVA

The repeated-measures ANOVA, in all of its forms, is used to test for mean differences in dependent samples. That is, it is akin to the dependent-samples t-test (see Chapter 5) in that a researcher collects data from the same participants more than once. This collection can allow a researcher to examine change over time within a single group of participants. The research questions that are most commonly answered with the one-way repeated measures ANOVA focus on the impact of time or treatment had on a group of individuals.

As with the ANOVAs discussed so far, repeated-measures ANOVA is used by researchers to test differences between group means. However, repeated-measures ANOVA is used when *all* members of a group are measured under different conditions. The reason we cannot use a standard ANOVA for this measurement is similar to the need for a dependent-samples t-test rather than an independent-samples t-test. In this case, the repeated-measures ANOVA incorporates the correlation between the repeated measures. Therefore, the data violates the standard ANOVA assumption of independence of observations.

As seen in Table 1.6 in Chapter 1, when researchers use a repeated-measures ANOVA, they are trying to answer a single research question: does a difference exist between three or more observations of the same phenomenon (paired sample) on one dependent variable? Note that I am specifically stating three or more observations. Yes, you could use a repeated-measures ANOVA to examine two observations of the same group. However, if only two observations exist, the more appropriate test would be a dependent-samples t-test.

For example, a music education researcher may be interested in following the intonations skills of his students over the course of a school year. To achieve this goal, he gives an intonation test at the beginning of the year, in the middle of the year, and at the end of the year. With this information, the researcher would be able to see if the music students improved their intonation skills over the course of the academic year to any significant level.

You should be clear about the difference between a repeated-measures design and a multivariate design. For both, sample members are measured on several occasions, or trials (in the multivariate design, these members would be separate dependent variables), but in the repeated-measures design, each trial represents the measurement of the same dependent variable under a different condition, after a different treatment, or at a different time. For example, you can use a repeated-measures ANOVA to compare the number of oranges produced by an orange grove at years one, two, and three. The measurement is the number of oranges, and the condition that changes is the year. In contrast, for the multivariate design, each trial represents the measurement of a different characteristic. You should not, for example, use a repeated-measures ANOVA to compare the number, weight, and price of oranges produced by a grove of orange trees. The three measurements are number, weight, and price, which do not represent different conditions, but different qualities. It is generally inappropriate to test for mean differences between such disparate measurements.

A slightly confusing aspect of the repeated-measures ANOVA, as well as distinguishing it from a multivariate design, is understanding what the independent and dependent variables are. As with a one-way ANOVA, a one-way repeated-measures ANOVA requires one continuous dependent variable and one categorical independent variable. However, the independent variable in a repeated-measures ANOVA is, more often than not, the different data collection points themselves. Let's contemplate our musical example from above in regard to a researcher who wanted to see if student scores on an intonation test had risen from the beginning of the year to the end. The researcher may have collected data from the same students at the beginning of the year, in the middle of the year, and at the end of the year. Because the researcher collected more than two points of data (or level), we cannot compute a dependent-samples t-test. In order to find out if any differences existed over time, the researcher would need to use a repeated-measures ANOVA. In this instance, in this example, the continuous dependent variable would be the scores on the intonation tests and the categorical independent variable would be the time of year (beginning, middle, end). In the published example found in Example 10.1, a researcher used a repeated-measures ANOVA to examine the impact of multiple listening opportunities on the error-detection accuracy of undergraduate music education majors (see Example 10.1).

Similarly, researchers may use the repeated-measures ANOVA to examine the impact of some treatment (or different treatments given to the same participants) over time. In Example 10.2, a researcher used a one-way ANOVA to examine the impact of three sequential training sessions on participants'

EXAMPLE 10.1 REPORTING OF A ONE-WAY REPEATED-MEASURES ANOVA

A repeated-measures analysis of variance (ANOVA) revealed significant differences in total correct error-identification responses across three listenings, $F(2, 150) = 161.98$, $p < .001$. On average, listeners correctly identified the greatest number of errors during the first listening ($M_1 = 28.59$, $SD = 9.96$), significantly fewer during the second listening ($M_2 = 14.04$, $SD = 7.99$), and the fewest during the third listening ($M_3 = 8.97$, $SD = 7.78$). Similar analysis revealed significant differences in total incorrect error-identification responses across three listenings as well, $F(2, 150) = 42.32$, $p < .001$. On average, listeners incorrectly identified placement of errors most often during the first listening ($M_1 = 30.29$, $SD = 13.48$), fewer during the second listening ($M_2 = 22.26$, $SD = 14.48$), and the fewest during the third listening ($M_3 = 18.25$, $SD = 13.54$).

Sheldon, D. A. (2004). Effects of multiple listenings on error-detection acuity in multivoice, multitimbral musical examples. *Journal of Research in Music Education, 52*(2), 102–115.

EXAMPLE 10.2 REPEATED-MEASURES DESIGN REPORTING WITH EFFECT SIZE

Data were examined by using a one-way, repeated-measures analysis of variance (ANOVA) and eta-squared effect size analysis. Post hoc comparisons were conducted as needed in the form of pairwise contrasts between the baseline and each training period mean.

Time-sampled skills in this category included positive affect, positive verbal reinforcement, and modeling of physical movements. See Table 1 for means and standard deviations. Positive affect increased significantly over time, $F(3, 12) = 18.2$, $p < .001$; $\eta^2 = .82$]. Post hoc comparisons located differences between the baseline and all three training periods: between baseline and Training 1, $F(1, 14) = 13.16$, $p = .003$; for baseline and Training 2, $F(1, 14) = 18.53$, $p = .001$; and from baseline to Training 3, $F(1, 14) = 63.42$, $p < .001$. Positive verbal reinforcement increased from baseline to Training 3, however, not significantly, $F(3, 12) = .55$, $p = .656$; $\eta^2 = .12$].

de l'Etoile, S. K. (2001). An in-service training program in music for child-care personnel working with infants and toddlers. *Journal of Research in Music Education, 49*(1), 6.

musical instruction for toddlers. In this example, the author was careful to report the effect size via eta squared.

Assumptions of the Repeated-Measures ANOVA

The repeated-measures ANOVA has many of the same assumptions as the ANOVAs discussed in previous chapters. It is different, however, in that the repeated-measures ANOVA does not require independent observations. It is required to collect data from the same person more than once, ensuring that data and subsequent analysis will be dependent upon the same person's score on the other observations. The assumptions of the repeated-measures ANOVA are the following:

1. The data collected for the dependent variable is continuous rather than categorical.
2. The data collected for the independent variable is categorical rather than continuous.
3. Normality—the dependent variable is normally distributed.
4. Homogeneity of variance—if there are multiple groups being examined, this term means that the groups all have similar variances, given the same dependent variable.
5. Sphericity—the variances of the repeated measures are all equal, and the correlations among the repeated measures are all equal. This step is required to test to see if we can compare the differences between the within-subjects (dependent) measures, which are most commonly tested by using the Mauchly's sphericity test (more to come on this).

Meeting the Assumptions of the Repeated-Measures ANOVA

Like many of the ANOVAs discussed so far, the repeated-measures ANOVA is robust against a few of the assumptions. However, it is still important to ensure that we use appropriate statistical tests with data that meets the assumptions so that our conclusions can be based upon defensible analyses. Taking care to use only categorical independent variables and continuous dependent variables is an easy first step to ensuring a good repeated-measures analysis. Although it is rather difficult to explicitly offer data on for this assumption, the normality can be checked by using the measures of central tendency and dispersion (mean, median, mode, and standard deviation) as well as kurtosis and skewness indices (remember that no skew and perfect kurtosis indices are 0). The sphericity assumption most often requires a Mauchly's sphericity test.

The Mauchly's sphericity test works as the Box M test works for the multivariate ANOVAs or the Levene test in a univariate analysis. Researchers use the Mauchly's sphericity test to examine whether or not a spherical matrix has equal variances and covariances equal to zero. The common covariance matrix of the transformed within-subject variables must be spherical, or the subsequent F tests and associated analyses and conclusions for whether or not to accept or reject the null hypothesis are invalid. As with the Levene test, as well as the Box M test, if the chi-square outcome of the Mauchly's test is less than .05, the analysis has not met the assumptions required for the repeated-measures ANOVA.

Options if You Do Not Meet the Assumption of Sphericity

A researcher has several options to consider if her data does not meet the assumption of sphericity as evidenced in the Mauchly's test for sphericity. The first option is to stop the analysis and either collect more data, find an alternative analysis option, or statistically alter the findings to account for the lack of sphericity. If you elect to try to account for the lack of sphericity, the two most common options are (1) the Greenhouse-Geisser adjustment, which is a relatively easy alteration to make but can over-correct and increase Type II error too much. Another, more moderate option is the Huynh-Feldt adjustment. However, some statisticians argue that Huynh-Feldt adjustments are too liberal and can increase a Type I error likelihood. In the end, it is up to the researcher to examine the type of research being conducted and whether or not possibly losing data or gaining too much data is more appropriate. Often, researchers make this decision on whether or not the research is exploratory or confirmatory. If a researcher is examining something early in a research agenda without an abundance of data from earlier studies, it might be more appropriate to err on the side of keeping more information and allowing more Type I errors. If, however, the researcher is trying to confirm information that has a large background of previous data, being more conservative (i.e., being more willing to commit a Type II error) is the more appropriate strategy. Luckily, SPSS (as well as most other statistical programs) automatically computes these alterations for you so you can decide for yourself. Regardless of what decision a researcher makes, he should always report the alteration and the reason for the selected change.

Research Design and the Repeated-Measures ANOVA

The repeated-measures ANOVA can be used in any research design in which the researcher has collected data from one group multiple times. This analysis is most often associated with any time series research design:

(A) O —— O —— O

or

(B) O —— O —— O —— X —— O —— O —— O

or, indeed, any variation of a time series design of one group with any given observations and treatments.

Let's take a look at a few musical examples of what these designs may look like.

(A) $O_{\text{start of year intonation score}} - O_{\text{middle of year intonation score}} - O_{\text{end of year intonation score}}$

In this example, as we have already discussed, a researcher was interested in whether or not participants' scores on an intonation test would improve over the course of the year. No specific treatment is applied (save for the treatments of time, maturation, etc.).

(B) $O_{rr} - O_{rr} - O_{rr} - X_{\text{reading unit}} - O_{rr} - O_{rr} - O_{rr}$

In this example, a music education researcher was interested in whether or not a unit of instruction on rhythm reading impacted students' rhythm reading (rr) skills. To examine this, the researcher tested the rhythm reading skills of students three times prior to the instructional unit and three times following. This design has some interesting benefits. Testing so many times allows students to get more comfortable with the test, allowing for a more typical performance. Moreover, the researchers will be better able to discuss the impact of maturation and testing effect on the outcome. Additionally, testing several times following the treatment allows the researcher to examine how lasting any impact of the treatment was as well as contextualize the outcomes in a broader manner.

The Null Hypotheses for the Repeated-Measures ANOVA

Researchers can use a repeated-measures ANOVA to test the null hypothesis that means for the group examined will not be different at different observations. The null hypothesis for this test can be visualized as

$H_0: \mu_{\text{1st observation}} = \mu_{\text{2nd observation}} = \mu_{\text{3rd observation}} = \mu_{\text{4th observation}} = \mu_5$ etc.

Keep in mind that we are looking at the means of the same group in this null hypothesis. The null hypothesis for our musical example could be written as

$H_0: \mu_{\text{start of year intonation}} = \mu_{\text{mid year intonation}} = \mu_{\text{end of year intonation}}$

The alternative hypotheses for the repeated-measures ANOVA would be

$H_a: \mu_{\text{1st observation}} \neq \mu_{\text{2nd observation}} \neq \mu_{\text{3rd observation}} \neq \mu_{\text{4th observation}} \neq \mu_5$ etc.

The alternative hypothesis for our musical example would be

$H_a: \mu_{\text{start of year intonation}} \neq \mu_{\text{mid year intonation}} \neq \mu_{\text{end of year intonation}}$

Finally, the directional hypothesis for a repeated-measures ANOVA would be:

$H_1: \mu_{1st\,observation} < \mu_{2nd\,observation} < \mu_{3rd\,observation} < \mu_{4th\,observation} < \mu_5$ etc.

If we had a strong reason to believe that the intonation scores of our students would consistently and significantly increase over the course of the school year, the directional hypothesis for our musical example would be

$H_a: \mu_{start\,of\,year\,intoation} < \mu_{mid\,year\,intoation} < \mu_{end\,of\,year\,intoation}$

Setting Up a Database for a Repeated-Measures ANOVA

The database for a repeated-measures ANOVA should generally have at least three columns. If you had only two columns, it would most likely be more appropriate to conduct a dependent-samples t-test, as discussed above. The three or more columns should contain information from the same participants (each row should have data from one individual), while each column should represent a different condition or time that the participants were observed or measured. In Figure 10.1, we can see a database in which each participant was given an intonation test at the beginning of a school year, in the middle of the school year, and at the end of the school year.

Conducting a Repeated-Measures ANOVA in SPSS

For this next section, please use the provided example dataset, Repeated Measures ANOVA Example 1 ▶, in order to complete the analysis as you

	EarlyYear	MidYear	EndYear
1	2	3	3
2	3	4	4
3	2	5	5
4	3	4	4
5	2	4	5
6	1	5	4
7	2	3	4
8	3	4	3
9	2	5	4
10	2	5	4
11	3	4	5
12	1	4	5
13	2	5	5
14	3	3	4
15	1	4	4

FIGURE 10.1

read. Although conducting a repeated-measures ANOVA in SPSS is relatively simple, and requires a few more steps than previous examples, but it can be achieved by following these steps:

1. Click **Analyze**.
2. Click **General Linear Model** (not **Generalized Linear Model**).
3. Click **Repeated Measures**. Once you have clicked **Repeated Measures**, you will get the screen shown in Figure 10.2.
4. Create a name for the within-subject factor you would like to create. As our example is about intonation, the word intonation would make sense to enter in **Within-Factor Subject Name**.
5. Enter the number of levels associated with the factor. In this case, as we observed participants three times, we have three levels.
6. Click **Add**.
7. Click **Define**. You will get the screen seen in Figure 10.3.
8. Move the three variables over to the **Within-Subjects Variables** *box in the appropriate order. In the example above, two of the three items have been moved over.*
9. Click **Plots**.
10. In the next screen, move the variable intonation into the **Horizontal Axis** *box and click* **Add**.
11. Click **Continue**.
12. Click **Options**. *You will be given the screen shown in Figure 10.4.*
13. As seen in the example above, move the factor (in our example, **intonation***) into the* **Display Means for** *box.*

FIGURE 10.2

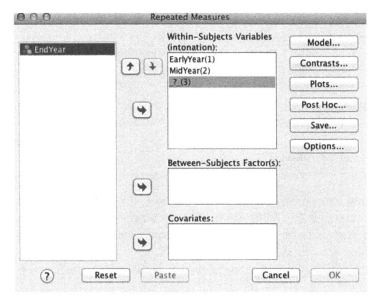

FIGURE 10.3

FIGURE 10.4

14. *Make sure that just below that box, you select* **Compare main effects**, *as shown.*
15. *Select* **Bonferroni** *from the* **Confidence interval adjustment** *drop-down menu just below, as shown.*
16. *Make sure that under* **Display**, *you select the boxes for* **Descriptive statistics** *and* **Estimates of effect sizes**. *Note that we do not need to select the* **Homogeneity tests**, *as we are not dealing with a between-subject design (at this point).*
17. *Click* **Continue**.
18. *Click* **OK** *to run the repeated-measures ANOVA.*

Once you have clicked **OK**, you will be presented with a number of boxes of information from SPSS. We will be focusing on only a few of them.

The first box is just a reminder of what we selected as the different levels of our within-subjects variable. See Figure 10.5

The box in Figure 10.5 shows us that we used early-year assessment, mid-year assessment, and an end-of-year assessment of intonation as our three levels.

The next important box is the descriptive statistics box (see Figure 10.6).

This box offers the researcher information about the means and standard deviations of all of the observations taken in the time-series design. It also lets us know if any data points are missing. In this instance, we see that we have data for 120 participants in each case. Therefore, we have no missing data. We can also see that our standard deviations are relatively similar.

The next box p of interest in our repeated-measures ANOVA is the Mauchly's test for sphericity (see Figure 10.7).

Within-Subjects Factors
Measure: MEASURE_1

Intonation	Dependent Variable
1	Early Year
2	Mid Year
3	End Year

FIGURE 10.5

Descriptive Statistics

	Mean	Std. Deviation	N
Early Year	2.08	.740	120
Mid Year	4.20	.729	120
End Year	4.22	.704	120

FIGURE 10.6

Mauchly's Test of Sphericity[a]

Measure: MEASURE_1

Within Subjects Effect	Mauchly's W	Approx. Chi-Square	df	Sig.	Epsilon[b]		
					Green house-Geisser	Huynh-Feldt	Lower-bound
intonation	.998	.244	2	.885	.998	1.000	.500

Tests the null hypothesis that the error covariance matrix of the orthonormalized transformed dependent variables is proportional to an identity matrix.

FIGURE 10.7

In this box, we see a Mauchly's W statistic (in this case .998) as well as the chi-square (.244) and related significance (.885). As with the Levene test as well as the Box M test, we see that our significance is over .05. Therefore, our groups are not significantly different, and we have met the assumption of sphericity. Neither the Greenhouse-Geisser nor Huynh-Feldt alterations are needed in this example.

The next informative box is the test of within-subjects effects (see Figure 10.8). This box tells us if any significant difference existed between observations.

In this example, we can assume sphericity (as evidenced by the not significant Mauchly's finding). Therefore, under **Source** and opposite **Intonation**, we follow the row **Sphericity Assumed**. If we had required an alteration, we could have used the more conservative Greenhouse-Geisser or the more liberal Huynh-Feldt alterations. We can see that the information provided looks very similar to an ANOVA table. There is good reason for that! This data provided is the same and it is reported just like the ANOVAs from past chapters. We can see that our F is large, our significance is less than .001, and our effect size is large as well (partial η^2 = .738).

Is this enough information? Not just yet. As in previous chapters, we need some type of post hoc test to tell us where the differences actually existed, as we had more than two observations. In this instance, however, we have already asked SPSS to compute the post hoc tests. When we asked SPSS to compute the main effects, we asked it to compute the differences between each group. Scroll down in the SPSS output until you see the box **Pairwise Comparisons**. This box (see Figure 10.9) is the post hoc test that will tell you where the differences existed.

From this table, we can see that the significant differences existed between observation 1 and 2 (beginning of the year to midyear), but not between observations 2 and 3 (midyear to end of the year). The researcher may conclude that intonation improves significantly in the early stages of instrumental tuition and then, for some reason as yet unexplained, slows.

Tests of Within-Subjects Effects

Measure: MEASURE_1

Source		Type III Sum of Squares	df	Mean Square	F	Sig.	Partial Eta Squared
intonation	Sphericity Assumed	362.706	2	181.353	335.557	.000	.738
	Greenhouse-Geisser	362.706	1.996	181.727	335.557	.000	.738
	Huynh-Feldt	362.706	2.000	181.353	335.557	.000	.738
	Lower-bound	362.706	1.000	362.706	335.557	.000	.738
Error (intonation)	Sphericity Assumed	128.628	238	.540			
	Greenhouse-Geisser	128.628	237.509	.542			
	Huynh-Feldt	128.628	238.000	.540			
	Lower-bound	128.628	119.000	1.081			

FIGURE 10.8

Pairwise Comparisons

Measure: MEASURE_1

(I) Intonation	(J) Intonation	Mean Difference (I−J)	Std. Error	Sig.[b]	95% Confidence Interval for Difference[b]	
					Lower Bound	Upper Bound
1	2	−2.117*	.096		−2.349	−1.885
	3	−2.142*	.096	.000	−2.376	−1.908
2	1	2.117*	.096	.000	1.885	2.349
	3	−.025	.093	1.000	−.250	.200
3	1	2.142*	.096	.000	1.908	2.376
	2	.025	.093	1.000	−.200	.250

Based on estimated marginal means
*The mean difference is significant at the
[b]Adjustment for multiple comparisons: Bonferroni.

FIGURE 10.9

The last interesting piece of information that is usually informative to the researcher, but not often included in a report, is the plot of the findings. See Figure 10.10.

From this plot, it is easy to see where the major change happened. Plots such as these are often more interesting with a greater number of observations, as it becomes possible to see trends in data over a long period of time.

Is this enough information? Yes. From the analysis thus far, you can report all of the descriptive statistics, the Mauchly's test outcome, the significance of the within-subjects test, and the effect size (partial eta squared) that was calculated for you in SPSS. You can also report the post hoc test findings and clarify them by using the descriptive statistics (mean and standard deviation).

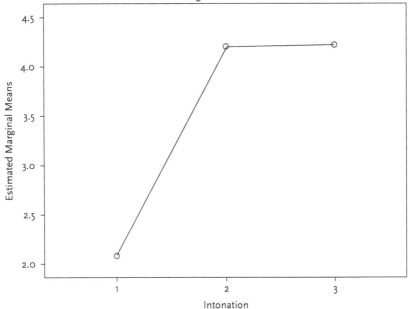

FIGURE 10.10

Reporting a Repeated-Measures ANOVA

Once you have conducted the repeated-measures ANOVA by using the provided dataset for this chapter, the final step is to create a research report that will provide the reader with all of the information necessary to make an informed decision about your research. Given the data set in our running example, one possible report of the data may look like Example 10.3.

Writing Hints for Repeated-Measures ANOVA

From the research report in Example 10.3, we see some hints that might help you state your findings as clearly as possible while providing all of the necessary information.

- Make sure that you provide the following in your report:
  o Mauchly's test for the sphericity assumption
  o N
  o F
  o df
  o p
  o η^2 for any significant finding
  o Discussion of practical and statistical significance

- o Means and standard deviations of all observations discussed
- o Post hoc tests you used and why
- Do not repeat the means of groups more than once. One you have reported it, you do not need to do so again.
- Another way to report the means and standard deviations of the groups could have been in a table.
- Per American Psychological Association guidelines, use first-person pronouns for clarity, especially when describing actions you took, while keeping all of your actions in past tense.
- Use the active voice.
- Avoid personification.
- Make sure all statistical indicators are in italics.
- Keep your actions as researcher in the past tense.

Repeated-Measures ANOVA Exercises

Given the second provided dataset, Repeated Measures ANOVA Example 2 ⊙, compute the most logical analysis. In this example, the columns are

- Rhythm Score at the beginning of a semester
- Rhythm Score after a quarter of the semester
- Rhythm Score at the middle of a semester
- Rhythm Score following a treatment of specialized rhythm instruction
- Rhythm Score at the end of the semester (still after the treatment)

EXAMPLE 10.3 POSSIBLE WRITE-UP OF A REPEATED-MEASURES ANOVA

In order to examine the intonation accuracy development of beginning instrumental music students, I administered a researcher-created intonation test to 120 participants at the beginning of the school year, in the midyear, and at the end of the school year. I employed a repeated-measures ANOVA to examine changes in intonation scores over the year. Then I established that the assumption of sphericity was met through a Mauchly's test (Mauchly = .998, χ^2 = .24, p = .89) and found a statistically significant difference between means of the different observations (F = 335.56, df = 2, p < .001, partial η^2 = .74) with a large effect size suggesting practical significance. Using post hoc tests, I found that a significant difference existed between the first observation (M_1 = 2.08, SD = .74) and the second (M_2 = 4.20, SD = .73) and third observation (M_3 = 4.23, SD = .70). Participants, however, showed no significant increase between the second and third observations.

The research design for this study would look like:

O —— O —— O —— X —— O —— O

This analysis might have you thinking like a teacher who is concerned about the long-term impact of instruction. Once you have found any significant changes, examine the plot to see where the changes exist. Begin to think about the implications of these findings and how you might explain them in a discussion section.

Once you have computed the most appropriate statistical tests, write a report that includes any descriptive statistics necessary to understand the repeated-measures ANOVA, as well as the test itself and, finally, effect size.

11 | Univariate Analysis of Covariance (ANCOVA)

Use of the Analysis of Covariance

As with the univariate ANOVA, music education researchers can use the univariate analysis of covariance (ANCOVA) when they need to compare the means of two or more groups or levels of a variable. Whereas the ANOVA examines the differences between means, however, the ANCOVA examines the differences between adjusted means.

As seen in Table 1.6 in Chapter 1, researchers using an ANCOVA are seeking to answer a simple question: do differences exist between two or more groups after controlling for one or more covariate (control variable) on one dependent variable?

When music education researchers use an ANCOVA to examine the differences between groups while controlling for a covariate, the independent variable (*must be categorical*) is the different groups or levels in the model, the covariate is a continuous variable that must be continuous and correlated with the dependent variable, and the dependent variable is a score (*must be continuous*) on some research instrument.

Before we go on, it may be beneficial to better understand what a covariate actually is:

In an ANCOVA a *covariate* is just another variable that is used to see if differences exist in the dependent variable. Rather than the usual categorical independent variables of the ANOVA, a covariate is a continuous variable that is correlated with the dependent variable. It is a means of controlling for a variable in order to better focus on the effect of the other independent variables on the dependent variable.

For example, a music education researcher may be interested in the physical discomfort experienced by collegiate instrumentalists when they play their instruments. Specifically, the researcher may be interested in the differences between those who play more than four hours a day and those who play less than four hours a day. We can begin to see the beginning of an ANOVA design (i.e., categorical independent variable of two groups, more than four hours, less than four hours, and one dependent variable, the reported scale of discomfort experienced while the participants were playing). However, what if the researcher also asked a question regarding how many minutes per day participants were physically active when not performing and found that this data was correlated to the reported discomfort while they were playing? The researcher would want to control for that variable to see if it changes the outcome of the analysis. In this example, the number of minutes of physical activity would be the covariate. Adjusting the data to account for the minutes of physical activity would help ensure that the analysis was more focused on the impact of the independent variable as it is controlling for (i.e., adjusting the means based on) the covariate of physical activity. That is, should a difference exist, the researcher would be in a better position to claim that the difference was due to the independent variable rather than the possibly confounding variable of other physical activity.

One of the most ubiquitous uses of an ANCOVA in music education research is to explore differences between groups in a pre-test/post-test design using the pre-test scores as the covariate. This is a different means of analyzing data from this type of design, as discussed in the chapters regarding ANOVA and t-tests, which focused on using the mean gain score as the dependent variable. See Example 11.1

Notice in Example 11.1, the researcher used the ANCOVA because the pre-test scores were significantly different among the three groups rather than correlated to the dependent variable (although given the design of the study, the pre-test scores were most likely correlated with the dependent variable). This practice helps to "level the playing field" and help focus the analysis on

---

EXAMPLE 11.1 USE OF ANCOVA IN A PRE-TEST/POST-TEST DESIGN.

In this example, primary analysis consisted of comparing means and standard deviations for all pre-test and post-test errors according to experimental conditions, as well as totals and means for each type of error across experimental groups. Because pre-test scores showed a significant difference among groups—$F(2, 66) = 4.377, p = .0164$]—analyses were conducted by ANCOVA with pre-test scores as the covariate. Results from this test revealed that there was no significant difference among groups because of treatment, $F = 2.71, p = .07$).

Kostka, M. J. (2000). The effects of error-detection practice on keyboard sight-reading achievement of undergraduate music majors. *Journal of Research in Music Education*, 48(2), 114–122.

> **EXAMPLE 11.2 USE OF ANCOVA TO CONTROL FOR POSSIBLE CONFOUNDING VARIABLE**
>
> In this study students in groups 1, 2, and 4 had more teachers while learning to play their main instrument (around 2.5) than did students in groups 3 and 5 (around 1.5). Analysis of covariance revealed that overall there was a significant difference among groups in the number of teachers for the main instrument even when the researchers were controlling for the number of years of playing: ANCOVA, $F(4, 249) = 2.89, p < .05$.
>
> ---
>
> Davidson, J. W., Moore, D. G., Sloboda, J. A., & Howe, M. J. (1998). Characteristics of music teachers and the progress of young instrumentalists. *Journal of Research in Music Education, 46*(1), 141–160.

the differences between the groups, while controlling for different levels of achievement not focused on the independent variable among the groups.

Not all uses of ANCOVA rely on data collected specifically in the design, but are used to control for possible confounding variables in a study. In Example 11.2, we can see researchers using this method. They have controlled for the number of years participants had been playing their instrument when examining the number of teachers with whom they have studied.

Post Hoc Tests for ANCOVA

As with the ANOVA, any ANCOVA analysis that examines the difference among more than two groups will require post hoc tests. The most common post hoc tests for an ANCOVA is the Bonferroni post hoc and is interpreted the same as the post hoc tests discussed in Chapter 8 (e.g., Tukey, Scheffe).

Assumptions of ANCOVA

The ANCOVA has many of the same assumptions as but is somewhat more complex than the univariate ANOVA. Therefore, the ANCOVA has a few more assumptions, as follow:

1. The data collected for the dependent variable is continuous rather than categorical (e.g., scores on a musical playing test).
2. The data collected for the independent variable is categorical (e.g., which instrument a student plays, which ensemble a student is in).
3. Each observation is independent of any other observation.
4. The dependent variable is normally distributed and has few or no outliers (see Chapter 1).

5. ANCOVA assumes homogeneity of variance, as discussed in previous chapters.
- The covariate should be correlated to the dependent variable.
- No interactions should exist between the covariate and the independent variable. This assumption is often referred to as homogeneity of regression slopes.

Meeting the Assumptions of the ANCOVA

Meeting all of the assumptions of an ANCOVA can be challenging. Several of the assumptions can be met simply by the design of the study (i.e., dependent variable is continuous, independent variable is categorical, covariate is continuous, and observations are independent). Meeting a few of the assumptions, however, is a little less clear. You can easily analyze having normal distribution and homogeneity of variance while you are conducting the ANCOVA. As we did in previous chapters regarding t-tests and ANOVAs, we can use the Levene test to examine homogeneity of variance, as we did for the t-tests. We can use simple bivariate correlation analyses to examine relationships between the variables and could even use a regression analysis to explore the homogeneity of regression slopes. However, this aspect might be more easily examined through looking at a graphical scatterplot of the data.

Research Design and the ANCOVA

The research designs most commonly employed with the ANCOVA are the same as those used with the ANOVA. The difference is that the means being examined are been adjusted on the basis of the covariate. Therefore, a few of the possible research designs that utilize this type of inferential test include the following (where O = an observation, X is a treatment of any given type):

(A) $O_{(adjusted\ mean)}$ X_1 $O_{(adjusted\ mean)}$

$O_{(adjusted\ mean)}$ X_2 $O_{(adjusted\ mean)}$

$O_{(adjusted\ mean)}$ X_3 $O_{(adjusted\ mean)}$

$O_{(adjusted\ mean)}$ X_4 $O_{(adjusted\ mean)}$

(B) X_1 $O_{(adjusted\ mean)}$

X_2 $O_{(adjusted\ mean)}$

$O_{(adjusted\ mean)}$

(C) $O_{(adjusted\ mean)}$
　　　―――――――――
　　$O_{(adjusted\ mean)}$
　　―――――――――
　　$O_{(adjusted\ mean)}$

Let's take a look at a few musical examples of what these designs might be in the area of music education.

(A) $O_{pre\text{-}discomfort\ (adjusted\ mean)}$ $X_{control\ group}$ $O_{post\text{-}discomfort(adjusted\ mean)}$
―――――――――――――――――――――――――――――――――――
　　$O_{pre\text{-}discomfort\ (adjusted\ mean)}$ $X_{1\ minute\ of\ warm\text{-}up}$ $O_{post\text{-}discomfort(adjusted\ mean)}$
―――――――――――――――――――――――――――――――――――
　　$O_{pre\text{-}discomfort\ (adjusted\ mean)}$ $X_{5\ minutes\ of\ warm\text{-}up}$ $O_{post\text{-}discomfort\ (adjusted\ mean)}$
―――――――――――――――――――――――――――――――――――
　　$O_{pre\text{-}discomfort\ (adjusted\ mean)}$ $X_{10\ minutes\ of\ warm\text{-}up}$ $O_{post\text{-}discomfort\ (adjusted\ mean)}$

In the example design above, a music education researcher was interested in the impact of different lengths of warm-up time on the physical discomfort reported by college instrumentalists. Participants in the control group did not warm up. The participants in the first treatment group warmed up for 1 minute while the second and third treatment group members warmed up for 5 and 10 minutes over the course of a semester. The researcher, however, found that scores on the pre-test were significantly different among groups. Therefore, the researcher used pre-test scores as a covariate in an ANCOVA to further explore the impact of warm-up duration on reported discomfort. The researcher could use an ANOVA to find out if any differences in post-test scores existed among the four groups while controlling for their initial self-report of physical discomfort. In this example, the students' score from post-test (final observation) serves as the dependent variable. The group that a participant is in serves as the independent variable, and the score on the pre-test (first observation) serves as the covariate.

As you can imagine, there are countless possible alterations to this design. There could be three groups only, or even more than four groups. Not all groups have to have a treatment.

If the researcher uses an ANCOVA in design B below, she can examine the differences among any number of groups without a pre-test while she is controlling for some other variable.

(B) $X_{1\ minute\ stretching}$ $O_{discomfort\ (adjusted\ mean)}$
―――――――――――――――――――――――
　　　$X_{5\ minutes\ stretching}$ $O_{discomfort\ (adjusted\ mean)}$
―――――――――――――――――――――――
　　　　　　　　　　　　　　$O_{discomfort\ (adjusted\ mean)}$

In this example (B), the same researcher was interested in the physical discomfort reported by instrumentalists, but was further interested in any differences among instrumentalists who took time to stretch prior to playing their instruments while she was controlling for the number of minutes active on an average day. This researcher would be able to use an ANCOVA to compare the discomfort ratings (dependent variable) of the participants in each group (independent variable) while controlling for the number of minutes active each day (covariate). This approach would help focus the analysis and conclusions on the actual differences between groups while controlling for the confounding variable.

In design C below, we see that no actual experimental design exists. Rather, researchers may use the ANCOVA to examine the differences among groups that are based upon individual difference variables (those that were not manipulated or controlled by the researcher (e.g., string players, wind players, percussionists).

(C) $O_{discomfort(adjusted\ mean)}$
_____

$O_{discomfort(adjusted\ mean)}$
_____

$O_{discomfort(adjusted\ mean)}$

In this example (C) the same researcher was interested in the physical discomfort reported by instrumentalists, but was further interested in any differences among string players, wind players, and percussionists. This researcher would be able to use an ANCOVA to compare the discomfort ratings (dependent variable) of the participants in each group (independent variable). Moreover, this researcher collected data on how many minutes a participant was active on an average day and found that it was correlated with his reported discomfort. Therefore, the researcher wanted to control for this variable and used the number of minutes active per day as a covariate. This approach would help focus the analysis and conclusions on the actual differences among the groups while controlling for the confounding variable.

The Null Hypothesis for the ANCOVA

The null hypothesis for the ANCOVA is that the means of all groups or levels of a variable will be equal when you are also controlling (i.e., adjusting means) for some covariate. It could be stated as

$H_0: \mu_{1\ (adjusted)} = \mu_{2\ (adjusted)} = \mu_{3\ (adjusted)} = \mu_{4\ (adjusted)} = \mu_{5\ (adjusted)}$ etc.

For our musical example above in regard to the impact of different lengths of warm-up activities, the null hypothesis would be that the means of the different treatment groups would be statistically the same.

$$H_0: \mu_{\text{control group(adjusted)}} = \mu_{1 \text{ minute of warm-up(adjusted)}} = \mu_{5 \text{ minutes of warm-up(adjusted)}} = \mu_{10 \text{ minutes of warm-up(adjusted)}}$$

The alternative hypothesis would be that there would be a difference between the mean of each of the groups or levels. It would be stated as

$$H_a: \mu_{1(\text{adjusted})} \neq \mu_{2(\text{adjusted})} \neq \mu_{3(\text{adjusted})} \neq \mu_{4(\text{adjusted})} \neq \mu_{5(\text{adjusted})} \text{ etc.}$$

In our musical example, the alternative hypothesis could be written as

$$H_a: \mu_{\text{control group(adjusted)}} \neq \mu_{1 \text{ minute of warm-up(adjusted)}} \neq \mu_{5 \text{ minutes of warm-up(adjusted)}} \neq \mu_{10 \text{ minutes of warm-up(adjusted)}}$$

Finally, the directional hypothesis would state that the mean of one or more of the groups or levels of variable would not be equal and that one would be higher or lower than the other. This hypothesis would be stated as

$$H_1: \mu_{1(\text{adjusted})} < \mu_{2(\text{adjusted})} < \mu_{3(\text{adjusted})} < \mu_{4(\text{adjusted})} < \mu_{5(\text{adjusted})} \text{ etc.}$$

In our musical example, the directional hypothesis might be listed as

$$H_1: \mu_{\text{control group(adjusted)}} < \mu_{1 \text{ minute of warm-up(adjusted)}} < \mu_{5 \text{ minutes of warm-up(adjusted)}} < \mu_{10 \text{ minutes of warm-up(adjusted)}}$$

As always, a researcher should be wary of directional hypotheses unless supported by evidence from previous researchers.

Setting Up a Database for the ANCOVA

The database for the univariate ANOVA will have at least *three* necessary columns. One column will have the dependent variable (some continuous variable). Another column will have the independent variable. The third column will contain the covariate variable. The independent variable should look something like the grouping variable of the ANOVA. In Figure 11.1, we can see a database established to compare the scores of four different groups of participants, which have been assigned a number (i.e., no warm-ups = 1, 1 minute of warm-ups = 2, 5 minutes of warm-ups = 3, 10 minutes of warm-ups = 4). The dependent variable is an overall self-reported physical discomfort score (scaled 0–10). The final column (covariate) is the self-reported number of minutes the participants are active on a typical day.

	Group	Discomfort	PhysicalActive
1	1	8	32
2	1	7	25
3	1	7	15
4	1	8	32
5	1	6	35
6	1	7	35
7	1	7	35
8	1	8	30
9	1	8	40
10	1	7	60
11	1	6	60
12	1	8	55
13	1	7	10
14	1	9	30
15	1	8	25

FIGURE 11.1

Therefore, the research design of this study would look like

$X_{\text{no stretching}}$ $O_{\text{discomfort (adjusted mean)}}$

$X_{\text{1 minute stretching}}$ $O_{\text{discomfort (adjusted mean)}}$

$X_{\text{5 minutes stretching}}$ $O_{\text{discomfort (adjusted mean)}}$

$X_{\text{10 minutes stretching}}$ $O_{\text{discomfort (adjusted mean)}}$

Conducting an ANCOVA in SPSS

For this section, please use the provided example database (ANCOVA Example 1 ▶) in order to go through the process while you are reading. This exercise will give you the guided experience of a step-by-step analysis so that you can conduct the follow-up analysis on the additional dataset on your own.

In order to run the ANCOVA, follow these steps:

1. Click **Analyze**.
2. Click **General Linear Model** (not **Generalized Linear Model**).
3. Click **Univariate**.
4. Enter the dependent variable in the **Dependent Variable** *slot*.
5. Enter the independent variable in the **Fixed Factor** *slot*.
6. Enter the covariate into the **Covariate** *slot*.
7. Click on **Options**.

8. *Select the independent variable (in this example **group**) and move it to the* **Display Means for** *slot.*
9. *Once you have moved the independent variable over, select* **Compare main effects** *and then select* **Bonferroni** *from the drop-down menu. This step will provide the post hoc pairwise comparisons should a significant outcome arise from the analysis.*
10. *In the* **Options** *screen make sure that you select* **Descriptive statistics, Homogeneity tests,** *and* **Estimates of effect sizes.**
11. *Click* **Continue.**
12. *Click* **OK.**

Once you have clicked **OK**, you should get five different boxes of information from SPSS that we will focus on: the descriptive statistics (not adjusted), homogeneity-of-error variance test, the test of between-subjects effects table (the actual ANCOVA outcome), the estimated means (aka the means adjusted for the covariate), and the pairwise comparisons.

In the first box, we simply see the descriptive statistics of each of the four groups as well as the aggregated total. See Figure 11.2.

From Figure 11.2, we can see that the participants in group 1 ($n = 40$) reported the highest level of discomfort ($M = 7.20$, $SD = .72$) followed by participants in group 2 ($n = 42$), who reported experiencing slightly less physical discomfort ($M = 7.00$, $SD = .66$). Participants in group 3 ($n = 37$) reported having experienced even less discomfort ($M = 6.14$, $SD = .82$), while those in group 4 ($n = 44$) reported the least amount of physical discomfort ($M = 4.41$, $SD = .76$).

The next box (see Figure 11.3) is the Levene test for homogeneity of error variance. Remember that, in this instance, *you do not want to have a significant difference.* You are looking for a significance level that is .05 or greater. If the number is smaller that .05, your groups do not have sufficiently similar variance to meet one of the fundamental assumptions of an ANCOVA (homogeneity of variance). We can see from Figure 11.3 that we did, indeed meet the assumption.

Descriptive Statistics
Dependent Variable: Discomfort

Group	Mean	Std. Deviation	N
1	7.20	.723	40
2	7.00	.663	42
3	6.14	.822	37
4	4.41	.757	44
Total	6.15	1.350	163

FIGURE 11.2

Levene's Test of Equality of Error Variances[a]

Dependent Variable: Discomfort

F	df1	df2	Sig.
1.756	3	159	.158

Tests the null hypothesis that the error variance of the dependent variable is equal across groups.

FIGURE 11.3

As the significance level (.158) is greater than .05, we can assume that the error variance between the groups (notice that the first degrees of freedom is calculated by groups ($N - 1$) and the second degrees of freedom is calculated on the basis of participants minus the number of groups ($N - 4$) is similar enough to meet the assumption of homogeneity of variance.

The next box created by SPSS that we will pay attention to is the table of the test of between-subjects effects, which contains the actual ANCOVA analysis. In this box (see Figure 11.4). we are most interested in the row that focuses on our independent variable. In this example, the independent variable has been labeled under **Source** as **Group**. Although more information is available (see Chapter 7 for a better description of the table), in this table, we are most interested in four pieces of information: F, df, p (sig), and partial eta squared (partial η^2). In this table we see that a significant difference does exist on the basis of what group participants were in and when researchers are controlling for the covariate ($F = 109.58$, $df = 3$, $p < .001$, partial $\eta^2 = .68$). Given the partial η^2 for this analysis (keep in mind, with only one predictor variable, partial η^2 and η^2 will be the same) it seems we can argue for some practical significance as well as statistical significance.

The next box (see Figure 11.5) is the estimated marginal means. This is the box that provides the means of the four groups once they have been adjusted on the basis of the covariate. We can see that the means have not drastically changed. The mean for group 1, for example, has gone down from 7.20 to 7.16. Note that not all changes will be in the same direction. Group 4, for example, started at 4.41 and increased to 4.43 once it was controlled for the covariate.

Following the estimated marginal means is the pairwise comparisons (see Figure 11.6). These are the post hoc tests we will use to see where the differences actually exist. As we see in Figure 11.6, statistical differences existed between groups 1 and 3, 1 and 4, 2 and 3, 2 and 4, and 3 and 4. We can use the mean-difference column of this table to interpret the findings, or we can look back to the estimated marginal means to interpret. Group 4 experienced less physical discomfort than any other group. Group 3 experienced significantly less discomfort that groups 1 and 2, but more than group 4.

Tests of Between-Subjects Effects
Dependent Variable: Discomfort

Source	Type III Sum of Squares	df	Mean Square	F	Sig.	Partial Eta Squared
Corrected Model	208.557[a]	4	52.139	95.117	.000	.707
Intercept	1270.742	1	1270.742	2318.206	.000	.936
Physical Active	.752	1	.752	1.371	.243	.009
Group	180.208	3	60.069	109.584	.000	.675
Error	86.609	158	.548			
Total	6467.000	163				
Corrected Total	295.166	162				

[a] R Squared = .707 (Adjusted R Squared = .699)

FIGURE 11.4

Estimates
Dependent Variable: Discomfort

Group	Mean	Std. Error	95% confidence Interval	
			Lower bound	Upper bound
1	7.157[a]	.123	6.915	7.399
2	6.983[a]	.115	6.756	7.211
3	6.171[a]	.125	5.923	6.419
4	4.434[a]	.114	4.210	4.659

[a] Covariates appearing in the model are evaluated at the following values: Physical Active = 56.56.

FIGURE 11.5

Pairwise Comparisons
Dependent Variable: Discomfort

(I) Group	(J) Group	Mean Difference (I-J)	Std. Error	Sig.[b]	95% Confidence Interval for Difference[b]	
					Lower Bound	Upper Bound
1	2	.174	.165	1.000	-.267	.615
	3	.986*	.182	.000	.501	1.472
	4	2.723*	.172	.000	2.264	3.182
2	1	-.174	.165	1.000	-.615	.267
	3	.812*	.173	.000	.350	1.274
	4	2.549*	.164	.000	2.112	2.986
3	1	-.986*	.182	.000	-1.472	-.501
	2	-.812*	.173	.000	-1.274	-.350
	4	1.737*	.165	.000	1.295	2.179
4	1	-2.723*	.172	.000	-3.182	-2.264
	2	-2.549*	.164	.000	-2.986	-2.112
	3	-1.737*	.165	.000	-2.179	-1.295

Based on estimated marginal means
*The mean difference is significant at the
[b] Adjustment for multiple comparisons: Bonferroni.

FIGURE 11.6

Reporting an ANCOVA

Once you have conducted the ANCOVA by using the provided dataset for this chapter, the final step is to create a research report that could be offered in a journal setting or dissertation writing. For readers not currently conducting their own research, looking at how these tests go from inception to reporting can give you a keen eye in determining the trustworthiness of the research report and the data contained within.

Given the grouping of the participants in our running music example in this chapter, one possible report of the data may look like Example 11.3.

EXAMPLE 11.3 POSSIBLE ANCOVA REPORT OF THE EXAMPLE DATASET

In order to examine the impact of warm-up duration on the self-reported physical discomfort experienced by instrumentalists, we utilized a post-test-only design. We randomly placed participants ($N = 163$) into one of four groups. Those in group 1 did not warm up prior to playing their instrument. Participants in group 2 warmed up for 1 minute, while those in group 3 warmed-up for 5 minutes. Participants in group 4 warmed up for 10 minutes. The treatment lasted for six weeks during a semester. Additionally, because we were curious about how the average number of minutes that participants were physically active during the day impacted physical discomfort, we asked participants for this information. We then conducted an ANCOVA with the treatment group as the independent variable, the participants' self-reported level of physical discomfort following the six weeks as the dependent variable, and number of minutes of physical activity in a typical day as the covariate.

Participants in group 1 ($n = 40$) reported the highest level of discomfort ($M = 7.16$, standard error, or SE, $= .12$) followed by participants in group 2 ($n = 42$), who reported experiencing slightly less physical discomfort ($M = 6.98$, $SE = .12$). Participants in group 3 ($n = 37$) reported having experienced even less discomfort ($M = 6.17$, $SE = .13$), while those in group 4 ($n = 44$) reported the least amount of physical discomfort ($M = 4.43$, $SE = .11$). We employed the ANCOVA and found a significant difference between groups ($F = 109.58$, $df = 3$, $p < .001$, partial $\eta^2 = .68$), as well as a practical significance based on the partial η^2. Through subsequent pairwise post hoc tests, we found that those who warmed up for 10 minutes experienced less physical discomfort than any other group, while those who warmed up for 5 minutes experienced significantly less discomfort than those who warmed up for 1 minute or not at all, but more than those who warmed up for 10 minutes. Effect sizes for the pairwise comparisons ranged between .35 and .54.

ANCOVA Writing Hints

From the report in Example 11.3, we see some hints that might help you state your findings as clearly as possible while providing all of the necessary information.

- Make sure that you provide the following in your report, as appropriate:
  - N
  - n
  - F
  - df
  - p
  - Partial η^2
  - Discussion of practical and statistical significance
  - Adjusted means and standard errors of all groups
  - Post hoc pairwise outcomes
- Do not repeat the means of groups more than once. One you have reported it, you do not need to do so again.
- Another way to report the adjusted means, standard errors, and pairwise comparisons of the groups may be in a table.
- Per American Psychological Association guidelines, use first-person pronouns for clarity, especially when describing actions you took, while keeping all of your actions in past tense.
- Use the active voice.
- Avoid personification (e.g., "data indicated that").
- Make sure all statistical indicators are in italics.
- Keep your actions as researcher in the past tense.

ANCOVA Exercises

Given the second provided dataset, ANCOVA Example 2 ▶, compute the most logical analysis. In this example, the three columns are the following:

- *Independent Variable: Grouping.* Group 1 includes high school orchestra students. Group 2 includes high school choral students. Group 3 includes high school band students.
- We found a significant difference between groups on the pre-test. Use it as the covariate.
- Use the post-test score as the dependent variable.

In this example dataset, a researcher was interested in whether or not differences existed in the rhythm reading skills of high school students in the different traditional ensembles (i.e., orchestra, choir, and band) after a few weeks of instruction on rhythm reading. Therefore, the researcher gave a pre-test and a post-test following the instruction. She found differences between the groups on the pre-test and then elected to use the pre-test scores as a covariate to control for them so she could see if the instruction had any impact on the post-test scores of students while she was controlling for the extant differences between the groups.

Once you have computed the most appropriate statistical test, write a report that includes any descriptive statistics necessary to understand the ANCOVA, as well as the test itself and, finally, effect size so that any reader can understand and trust your analysis.

12 | Multivariate Analysis of Covariance (MANCOVA)

Use of the Multivariate Analysis of Covariance (MANCOVA)

What is a researcher to do when he wants to create a statistical model in order to compare the means of more than one dependent variable for different groups while also controlling for some additional covariate? As we have seen so far, the MANOVA is the most likely statistical procedure for examining the means of two or more dependent variables when the independent variable is categorical. We have also learned in a previous chapter that a covariate is a continuous variable that a researcher would like to control for in her analysis. Therefore, it is a logical leap to make that the most appropriate test in this instance would be a multivariate analysis of covariance (MANCOVA).

As seen in Table 1.6 in Chapter 1, researchers using a MANCOVA are attempting to answer a relatively complex research question: do differences exist between two or more groups after controlling for one or more covariates on two or more dependent variables?

For example, a music education researcher may be examining the possible connections between three different types of sight-singing instruction (i.e., numbers, letter names, and solfège) to students to sing accurate pitches as well as rhythms. Knowing that students may come to this task with different levels of musical aptitude, the researcher would like to control for the differences by using some measure of musical aptitude. In this example, the independent variable is the type of sight-singing instruction given. The dependent variables are scores on two measures, one on pitch accuracy, the other on rhythm accuracy. The covariate is the students' score on some measure of musical aptitude.

In Example 12.1, we can see how music education researchers employed a MANOVA to examine the impact of two different musical interventions (one in which students actively composed and another in which students experienced a more traditional teacher-centered musical experience) on academic achievement in spelling, reading comprehension, and mathematics. They also wanted to control for some variable to focus the analysis on the two different interventions. Therefore, they collected data on participants' age and also included pre-test results as a way to account for learning extant before the treatment.

In Example 12.1, we see that the researchers did find a significant outcome on the omnibus test. Much as with the MANOVA, however, we still don't know where the differences actually existed. Whereas once researchers using a MANOVA found that the overall test was significant (the Wilks's Λ), they would need to conduct subsequent ANOVAs for each of the independent variables to find where significance existed, researchers employing a MANCOVA would need to conduct follow-up ANCOVAs (see Chapter 11). This follow-up is what the authors from the study in Example 12.2 did.

EXAMPLE 12.1 REPORTING OF MANCOVA

To address the third research question, concerning the effects of the intervention on academic achievement, researchers performed a MANCOVA with language (spelling), reading comprehension, and mathematics being dependent variables, and controlled for age and pre-test results, followed by three separate ANCOVAs. In the MANCOVA and ANCOVA models, all covariate pre-test scores were statistically significant, $p < .005$. The results of the MANCOVA tests with the use of Wilks's criterion indicates a marginally significant effect of condition on combined dependent variables, Wilks's $\lambda = .933$, $F(3, 104) = 2.48$, $p = .065$, partial $\eta^2 = .067$.

Hogenes, M., van Oers, B., Diekstra, R. F., & Sklad, M. (2015). The effects of music composition as a classroom activity on engagement in music education and academic and music achievement: A quasi-experimental study. *International Journal of Music Education*, 0255761415584296.

EXAMPLE 12.2 REPORTING OF FOLLOW-UP ANCOVAS FOR MANCOVA ANALYSIS

Follow-up separate ANCOVAs subsequent to Example 12.1 showed a significant effect of the condition on one out of three academic achievement indicators. The ANCOVA showed a statistically significant—$F(1, 112) = 7.42$, $p = .007$, partial $\eta^2 = .06$—effect of condition on reading comprehension measured at the post-test after adjustment for covariates: age and pre-test score of reading comprehension.

Hogenes, M., van Oers, B., Diekstra, R. F., & Sklad, M. (2015). The effects of music composition as a classroom activity on engagement in music education and academic and music achievement: A quasi-experimental study. *International Journal of Music Education*, 0255761415584296.

In this example, the independent variable (i.e., the two interventions) had only two levels. Therefore, no additional post hoc tests were required. If, however, any significant independent variables have more than two levels, researchers would need to conduct additional post hoc tests on the ANCOVA outcomes to see where the significant differences existed. See Chapter 11 for more information on how to conduct and interpret these post hoc tests.

MANCOVA Assumptions

The MANCOVA has many of the same assumptions as the MANOVA discussed in Chapter 9. One additional assumptions exists, however, due to having two or more dependent variables. The assumptions of the MANCOVA are the following:

1. The data collected for the dependent variables is continuous rather than categorical.
2. The data collected for the independent variable is categorical rather than continuous.
3. Each observation is independent of any other observation.
4. Multivariate normality— all of the dependent variables are normally distributed themselves and any combination of the dependent variables is normally distributed.
5. Homogeneity of the covariance matrices—each of the dependent variables has equal variance when compared to each independent variable. This equality is assessed by the use of a Box M test (see Chapter 9 on MANOVA for more information).
6. Linearity—there is a linear relationship between independent variables and dependent variables.
7. Absence of multicolinearity—two or more dependent variable should not be highly correlated to one another.

Meeting the Assumptions of MANCOVA

Although the MANCOVA is robust against a few of the assumptions, as is the MANOVA, it is still important to ensure that we use appropriate statistical tests with data that meet the assumptions so that our conclusions can be based upon defensible analyses. Many of the assumptions of the MANCOVA can be taken care of through the design of the study and types of data collected. Others are more easily examined during the actual data analysis. The homogeneity of the covariance matrices assumption, for example, can be examined through the Box M test (See Chapter 9 for more information on the Box M test). Music education researchers can use simply bivariate correlations to check for the absence of multicolinearity and simple scatterplots to check for linearity.

MANCOVA Research Design

The MANCOVA can be used in any research design that could also be used for the MANOVA. The only difference is that the means have been adjusted to account for or control for the covariate or covariates included in the analysis. The MANCOVA builds upon the ANCOVA, but can be used to create a slightly more complex analysis by examining the impact of one or more categorical independent variables on more than one dependent variable while still controlling for any covariate. A few possible research designs for the MANCOVA would be the following:

(A) $O_{\text{(adj. X dependent variable 1)}}$ $O_{\text{(adj. X dependent variable 2)}}$ X_1 $O_{\text{(adj. X dependent variable 1)}}$ $O_{\text{(adj. X dependent variable 2)}}$

$O_{\text{(adj. X dependent variable 1)}}$ $O_{\text{(adj. X dependent variable 2)}}$ X_2 $O_{\text{(adj. X dependent variable 1)}}$ $O_{\text{(adj. X dependent variable 2)}}$

(B) X_1 $O_{\text{(adj. X dependent variable 1)}}$ $O_{\text{(adj. X dependent variable 2)}}$

X_2 $O_{\text{(adj. X dependent variable 1)}}$ $O_{\text{(adj. X dependent variable 2)}}$

$O_{\text{(adj. X dependent variable 1)}}$ $O_{\text{(adj. X dependent variable 2)}}$

(C) $O_{\text{(adj. X dependent variable 1)}}$ $O_{\text{(adj. X dependent variable 2)}}$

$O_{\text{(adj. X dependent variable 1)}}$ $O_{\text{(adj. X dependent variable 2)}}$

$O_{\text{(adj. X dependent variable 1)}}$ $O_{\text{(adj. X dependent variable 2)}}$

Please keep in mind that these example have only two dependent variables, but a MANCOVA (and MANOVA) can be used with more than two dependent variables and more than one independent variable. The designs with more than one independent variable are sometimes referred to as factorial MANCOVAs (or MANOVAs).

Let's take a look at what a few music education research examples might be for these designs.

(A) $O_{\text{(adj. pre rhythm score)}}$ $O_{\text{(adj. pre pitch score)}}$ X_{numbers} $O_{\text{(adj. post rhythm score)}}$ $O_{\text{(adj. post pitch score)}}$

$O_{\text{(adj. pre rhythm score)}}$ $O_{\text{(adj. pre pitch score)}}$ $X_{\text{letter name}}$ $O_{\text{(adj. post rhythm score)}}$ $O_{\text{(adj. post pitch score)}}$

$O_{\text{(adj. pre rhythm score)}}$ $O_{\text{(adj. pre pitch score)}}$ $X_{\text{solfège}}$ $O_{\text{(adj. post rhythm score)}}$ $O_{\text{(adj. post pitch score)}}$

In this research design example, we see the design of the study discussed at the opening of this chapter. A music education researcher was interested in the impact, if any, of three different sight-singing techniques (use of numbers, letter names, and solfège) on participants' ability to sing accurate pitches as well as accurate rhythms. Additionally, the researcher wanted to account for preexisting musical aptitude and collected information on each student on some scaled measure of musical aptitude to use as a covariate. If the pre-tests were different between groups, the researcher might also use the pre-test scores as covariates to control for previous learning or achievement.

(B) $X_{\text{classical music}}$ $O_{\text{(adj. X enjoyment)}}$ $O_{\text{(adj. X anxiety)}}$

$X_{\text{popular music}}$ $O_{\text{(adj. X enjoyment)}}$ $O_{\text{(adj. X anxiety)}}$

$O_{\text{(adj. X enjoyment)}}$ $O_{\text{(adj. X anxiety)}}$

In the example above, the music researcher was interested in a couple of psychological outcomes based on what music people listened to for one month. In this study, three groups of participants either listened to nothing but classical music, nothing but popular music, or no music for at least two hours per day. Additionally, the researcher thought that age might play a role in the outcome of the study and collected that information. In this example, the grouping is the independent variable, the two dependent variables are two psychological tests, one regarding the level of enjoyment of music, and the other on the level of anxiety felt during music listening. The covariate for the study would be the reported age of each participant.

(C) $O_{\text{(adj. X nonwestern songs)}}$ $O_{\text{(adj. X dependent variable 2)}}$

$O_{\text{(adj. X nonwestern songs)}}$ $O_{\text{(adj. X dependent variable 2)}}$

$O_{\text{(adj. X nonwestern songs)}}$ $O_{\text{(adj. X dependent variable 2)}}$

In this example, we see that the researcher was simply interested in comparing three different groups on two dependent variables. This is the kind of design that is often used in descriptive studies or survey studies. In this example, a researcher conducted a survey of elementary music teachers and was curious as to how geographic setting impacted the number of nonwestern songs sung in class and the extent to which teachers valued culturally responsive teaching on the basis of a self-reported scale. The three different groups of elementary music teachers (i.e., East Coast teachers, teachers in the Midwest, and West Coast teachers) all responded to the same survey. Also in the survey was an item asking the participants about their years of experience teaching music. The researcher would be able to use the information about teaching experience to control for that variable.

The Null Hypotheses for the MANCOVA

Researchers can use a MANCOVA to test the hypothesis that one or more independent variables, or factors, have an effect on two or more dependent variables while controlling for some covariate or covariates. As with the MANOVA, the researcher is interested in both main effects and interactions. However, in addition to the null hypotheses stated for the MANOVA, the researcher is interested in how the adjusted means (based on the covariate) differ or not. The null hypothesis could be stated as

H_0: $\mu_{adj.1} = \mu_{adj.2} = \mu_{adj.3} = \mu_{adj.4}$ etc. for Dependent Variable 1
H_0: $\mu_{adj.1} = \mu_{adj.2} = \mu_{adj.3} = \mu_{adj.4}$ etc. for Dependent Variable 2
Etc.

For our musical example above in regard to scores on pitch and rhythm tests following three different sight-singing types of instruction and controlling for music aptitude, the null hypothesis could be stated as

H_0: $\mu_{adj.numbers} \neq \mu_{adj.letter names} \neq \mu_{adj.solfège}$ for Pitch and for Rhythm

The alternative hypotheses for the MANOVA would be

H_a: $\mu_{adj.1} \neq \mu_{adj.2} \neq \mu_{adj.3} \neq \mu_{adj.4}$ ... etc. for Dependent Variable 1
H_a: $\mu_{adj.1} \neq \mu_{adj.2} \neq \mu_{adj.3} \neq \mu_{adj.4}$ ... etc. for Dependent Variable 2
Etc.

For our musical example, the alternative hypothesis could be stated as

H_a: $\mu_{adj.numbers} \neq \mu_{adj.letter names} \neq \mu_{adj.solfège}$ for Pitch and for Rhythm

Finally, the directional hypothesis could be written in a number of ways, depending on what the researcher already knows about the phenomenon being studied. All of the directional hypotheses could be written in the same direction, as can be most clearly seen in our musical example:

H_1: $\mu_{adj.numbers} < \mu_{adj.letter names} < \mu_{adj.solfège}$ for Pitch and for Rhythm

However, not all of the directional hypotheses need to be in the same direction. That is, given that there are multiple dependent variables, a researcher could hypothesize that one variable would go in one direction while the other would go in the other direction. For example.

H_1: $\mu_{adj.numbers} < \mu_{adj.letter names} < \mu_{adj.solfège}$ for Pitch

H_1: $\mu_{adj.numbers} > \mu_{adj.letter names} > \mu_{adj.solfège}$ for Rhythm

Setting Up a Database for a MANCOVA

The database for a MANCOVA should have at least four columns. It can have more, depending on the number of independent variables and dependent variables. For the purposes of this chapter, we will focus on a one-way MANCOVA, which means that we will be working with only one independent variable. A MANCOVA, however, can work with multiple independent variables. This approach is known as a factorial MANCOVA and works much like the principles found in Chapter 8. In our MANCOVA, at least two columns of continuous dependent variables are needed with at least one column of a categorical independent variable. The independent-variable column should look just like an independent-variable column you created for an ANOVA, factorial ANOVA, or MANOVA. The fourth column that is required for the MANCOVA is the one containing the covariate.

In Figure 12.1, we see a database created to run a MANCOVA with two dependent variables, one independent variable, and one covariate. In this database, the two dependent variables are the scores (on a scale of 1–10) on a pitch accuracy test and a rhythm accuracy test during participants' sight-reading. The independent variable is the type of sight-singing instruction they received (1 = numbers, 2 = letter names, 3 = solfège). The covariate is their score on a given musical aptitude test.

Therefore the research design for this database to be examined throughout this chapter would be

	Sightsingmethod	PostRhythm	PostPitch	Aptidtude
1	1	8	5	89
2	1	7	6	88
3	1	8	6	77
4	1	8	6	98
5	1	9	7	78
6	1	6	7	88
7	1	8	7	95
8	1	8	7	95
9	1	7	4	94
10	1	7	7	96
11	1	8	5	98
12	1	8	6	94
13	1	9	7	88
14	1	9	7	98
15	1	6	6	78

FIGURE 12.1

(A) $X_{numbers}$ $O_{(adj.\ post\ rhythm\ score)}$ $O_{(adj.\ post\ pitch\ score)}$

$X_{letter\ name}$ $O_{(adj.\ post\ rhythm\ score)}$ $O_{(adj.\ post\ pitch\ score)}$

$X_{solfege}$ $O_{(adj.\ post\ rhythm\ score)}$ $O_{(adj.\ post\ pitch\ score)}$

Conducting a MANCOVA in SPSS

For this section, please use the provided dataset, MANCOVA Example 1 ▶, to conduct the analysis as you read. Conducting a MANCOVA in SPSS will look rather similar to running an ANCOVA and can be accomplished by following these steps:

1. *Click* **Analyze**.
2. *Click* **General Linear Model** (not **Generalized Linear Model**).
3. *Click* **Multivariate** *(remember that we have more than one dependent variable)*.
4. *Highlight the dependent variables (in this example it is the* **Pitch and Rhythm** *scores) and move it into the* **Dependent Variable** *slot.*
5. *Highlight the independent variable (in this example* **Sightsingmethod***) and move it into the* **Fixed Factor(s)** *slot.*
6. *Highlight the covariate (in this example* **Aptitude***) and move it to the* **Covariate** *slot.*
7. *Click on* **Options** *and make sure that you have selected each of the following:*
   - **Descriptive statistics**
   - **Estimates of effect size**
   - **Homogeneity tests**
8. *Select the independent variable (***Sightsingmethod***) and move it into the* **Display Means for** *box.*
9. *Select the* **Compare main effects** *box.*
10. *From the drop-down menu under* **Confidence interval adjustment** *select* **Bonferroni**.
10. *Click* **Continue**.
11. *Click* **OK** *to run the MANOVA.*

Once you have clicked **OK**, you should get several boxes of information. We will sort through the most appropriate and meaningful here. The first box is just a listing of the number of participants in each group as broken down by the independent variable. See Figure 12.2.

The second box created by SPSS provides the same information, as found in Figure 12.2, but also provides the means and standard deviations, based on independent variable grouping for all (in this case both) dependent variables of each group (see Figure 12.3). From this box we can see that very little difference seems to exist between groups regarding rhythm (means range from 7.67 to 7.85), but more difference exists for the outcome of the pitch test (means range from 6.13 to 8.25).

The third box that SPSS creates gives the researcher the Box M test results (see Figure 12.4). As with the MANOVA, this test is analogous to the Levene test and tells the researcher if she has met the assumption of homogeneity of covariance matrices. This is one of the tests in which a significant outcome is less than ideal. If the differences of covariance matrices

Between-Subject Factors

		N
Sight sing method	1	48
	2	52
	3	51

FIGURE 12.2

Descriptive Statistics

	Sight sing method	Mean	Std. Deviation	N
Post Rhythm	1	7.75	.934	48
	2	7.85	1.092	52
	3	7.67	1.160	51
	Total	7.75	1.064	151
Post Pitch	1	6.13	.959	48
	2	7.00	1.085	52
	3	8.25	.935	51
	Total	7.15	1.319	151

FIGURE 12.3

Box's Test of Equality of Covariance Matrices[a]

Box's M	7.948
F	1.299
df1	6
df2	529143.345
Sig.	.254

Tests the null hypothesis that the observed covariance matrices of the dependent variables are equal across groups.

FIGURE 12.4

is greater than chance, we are no longer comparing apples to apples. We are hoping for a not significant outcome in the Box M test. In our example, it appears that we have met the assumption, as our significance level is greater than .05.

The next provided box gives us the actual outcome of the MANCOVA (see Figure 12.5). In this table, the researcher is provided with the Wilks's lambda and its corresponding F and significance as well as the partial eta squared. As with the MANOVA, SPSS offers four different means of evaluating the significance of the overall MANCOVA model: Pillai's trace, Wilks's lambda, Hotelling's trace, and Roy's largest root. Please review Chapter 9 for more information about deciding, on the basis of your analysis, upon which test to focus. For this example, we will continue focusing on Wilks's lambda.

In order to interpret the table in Figure 12.5, we focus on the row that indicates the independent variable (in this case **Sightsingmethod**). We see that the Wilks's lambda (λ) equals .576 with an F of 23.197, significance of < .001, and a partial eta squared of .241. The significant lambda tells us that the overall model is significant, meaning that we have somewhere found some difference based upon the independent variable. We must look to the test of between-subject outcomes in the next two boxes to know on the basis of which dependent variable or variables the differences exist.

Multivariate Tesis[a]

Effect		Value	F	Hypothesis df	Error df	Sig.	Partial Eta Squared
Intercept	Pillai's Trace	.324	34.978[b]	2.000	146.000	.000	.324
	Wilks' Lambda	.676	34.978[b]	2.000	146.000	.000	.324
	Hotelling's Trace	.479	34.978[b]	2.000	146.000	.000	.324
	Roy's Largest Root	.479	34.978[b]	2.000	146.000	.000	.324
Aptidtude	Pillai's Trace	.003	.212[b]	2.000	146.000	.809	.003
	Wilks' Lambda	.997	.212[b]	2.000	146.000	.809	.003
	Hotelling's Trace	.003	.212[b]	2.000	146.000	.809	.003
	Roy's Largest Root	.003	.212[b]	2.000	146.000	.809	.003
Sight sing method	Pillai's Trace	.426	19.873	4.000	294.000	.000	.213
	Wilks' Lambda	.576	23.197[b]	4.000	292.000	.000	.241
	Hotelling's Trace	.734	26.601	4.000	290.000	.000	.268
	Roy's Largest Root	.730	53.669[c]	2.000	147.000	.000	.422

[a]Design: Intercept + Aptidtude + Sight sing method
[b]Exact statistic

FIGURE 12.5

The fifth box created by SPSS will look very familiar to you. It is a Levene test of equality of error variance (see Figure 12.6). Remember that Box M told us that we met the assumption of the MANCOVA; this Levene test is there to tell us if we met the assumption of the post hoc ANOVAs that are automatically conducted within the MANCOVA.

From this box we can examine the equality of error variance of both dependent variables. Like the Box M and all the Levene tests we have discussed before, we are hoping for a not significant outcome on these tests and have achieved it here. We see that the Levene statistic for Rhythm is .228 and the statistic for Pitch is .955, both above .05. Had we violated this assumption, it would not have been the end of the analysis, but rather would have informed the type of post hoc tests we would have run to complete the analysis. See Chapter 11 for more information about this subject.

The next box that SPSS creates in this analysis is the test of between-subjects effects and is the post hoc ANCOVAs being conducted to see where the difference found in the overall model by our Wilks's lambda actually exist (see Figure 12.7). We see that the only statistically significant outcome based on the independent variable of the sight-singing method is on the dependent variable of pitch. We have an F of 53.39, a df of 2, a significant finding of $p < .001$, and a partial eta squared of .421. On the basis of that partial-eta-squared outcome, we also have a decent argument for practical significance as well.

The next box created by SPSS in this analysis is the estimated marginal means (see Figure 12.8) and provides the means of each group by dependent variable in their adjusted form. In our example, they have been adjusted on the basis of scores from an aptitude test. As you can see, although aptitude was significantly correlated with pitch, the means have changed relatively little from those stated in Figure 12.3.

We still do not quite have all of the information that we need. From our analysis so far, on the basis of the Wilks's lambda we know that something in the model is significant. From the follow-up ANOCVAs we know that the difference is found somewhere in regard to the pitch dependent variable. However, as we have three groups, we still need to know between which groups that

Levene's Test of Equality of Error Variances[a]

	F	df1	df2	Sig.
Post Rhythm	1.493	2	148	.228
Post Pitch	.046	2	148	.955

Tests the null hypothesis that the error variance of the dependent variable is equal across groups.
[a] Design: Intercept + Aptidtude + Sight sing method

FIGURE 12.6

Tests of Between-Subjects Effects

Source	Dependent Variable	Type III Sum of Squares	df	Mean Square	F	Sig.	Partial Eta Squared
Corrected Model	Post Rhythm	1.315a	3	.438	.382	.766	.008
	Post Pitch	113.868b	3	37.956	37.975	.000	.437
Intercept	Post Rhythm	39.013	1	39.013	34.011	.000	.188
	Post Pitch	39.423	1	39.423	39.443	.000	.212
Aptidtude	Post Rhythm	.483	1	.483	.421	.517	.003
	Post Pitch	.010	1	.010	.010	.921	.000
Sight sing method	Post Rhythm	.720	2	.360	.314	.731	.004
	Post Pitch	106.727	2	53.364	53.390	.000	.421
Error	Post Rhythm	168.619	147	1.147			
	Post Pitch	146.926	147	.999			
Total	Post Rhythm	9251.000	151				
	Post Pitch	7971.000	151				
Corrected Total	Post Rhythm	169.934	150				
	Post Pitch	260.795	150				

aR Squared = .008 (Adjusted R Squared = −.013)
bR Squared = .437 (Adjusted R Squared = .425)

FIGURE 12.7

Estimates

Dependent Variable	Sight sing method	Mean	Std. Error	95% Confidence Interval	
				Lower Bound	Upper Bound
Post Rhythm	1	7.730a	.158	7.418	8.041
	2	7.847a	.149	7.553	8.141
	3	7.685a	.153	7.383	7.986
Post Pitch	1	6.122a	.147	5.831	6.413
	2	7.000a	.139	6.726	7.274
	3	8.257a	.142	7.976	8.539

aCovariates appearing in the model are evaluated at the following values: Aptidtude = 89.27.

FIGURE 12.8

difference exists. We will find this out in the next box created in SPSS. The box (see Figure 12.9) contains the pairwise comparisons we asked for with the Bonferroni adjustment we set as well .

From Figure 12.9, we can see that all three groups are significantly different regarding the Pitch test even while they are controlled for participants' musical aptitude. As with the ANCOVA, you can contextualize these findings either with the adjusted means or by using the mean difference column in Figure 12.9.

Pairwise Comparisons

Dependent Variable	(I) Sight sing method	(J) Sight sing method	Mean Difference (I−J)	Std. Error	Sig.[b]	95% Confidence Interval for Difference[b]	
						Lower Bound	Upper Bound
Post Rythm	1	2	−.117	.217	1.000	−.642	.408
		3	.045	.223	1.000	−.495	.586
	2	1	.117	.217	1.000	−.408	.642
		3	.162	.213	1.000	−.353	.677
	3	1	−.045	.223	1.000	−.586	.495
		2	−.162	.213	1.000	−.677	−.353
Post Pitch	1	2	−.878*	.202	.000	−1.368	−.388
		3	−2.135	.208	.000	−2.640	−1.631
	2	1	.878*	.202	.000	.388	1.368
		3	−1.257*	.199	.000	−1.738	−.776
	3	1	2.135*	.208	.000	1.631	2.640
		2	1.257*	.199	.000	.776	1.738

Based on estimated marginal means
*The mean difference is significant at the
[b]Adjustment for multiple comparisons: Bonferroni.

FIGURE 12.9

Reporting a MANCOVA

Now that we have all of the information we need from the analysis, we can begin to consider how to best report the information. Given the data from the analysis discussed in this chapter, one possible report of the data may look like the one found in Example 12.3.

MANCOVA Writing Hints

From the research report in Example 12.3, we see some hints that might help you state your findings as clearly as possible while providing all of the necessary information for readers to understand and trust your analysis.

- Make sure that you provide the following in your report:
  - Lambda and its corresponding F, significance, and effect size
  - Box M for the MANOVA assumption
  - Levene finding for each univariate follow-up test
  - N
  - n
  - F for each test (including Levene)
  - df for each test
  - p for each comparison
  - η^2 for each significant finding
  - Discussion of practical and statistical significance
  - Adjusted means and standard errors of all groups discussed
  - Post hoc tests you used and why
- Do not repeat the means of groups more than once. One you have reported it, you do not need to do so again.
- Another way to report some of the descriptive data is in a table. This approach would be appropriate for the inferential analysis as well, especially if more variables are added to the analysis.
- Per American Psychological Association guidelines, use first-person pronouns for clarity, especially when describing actions you took, while keeping all of your actions in past tense.
- Use the active voice.
- Avoid personification.
- Make sure all statistical indicators are in italics.
- Keep your actions as researcher in the past tense.

MANCOVA Exercises

Given the second provided dataset for this chapter, MACNOVA Example 2 ▶, compute the most logical analysis. In this example, the columns are the following:

EXAMPLE 12.3 POSSIBLE MANCOVA REPORT OF THE EXAMPLE DATASET

In order to explore to the impact of three different types of sight-singing approaches on first-year university music students' ($N = 151$) ability to sight-sing accurate pitches and rhythms, I employed a MANCOVA analysis. In this analysis, participants' scores on two different assessments, one focused on pitch and one focused on rhythm, served as the dependent variables. The independent variable was the type of sight-singing approach students used in their class. Some students learned by using a number system for pitches ($n = 48$), some used note letter names ($n = 52$), while some used solfège ($n = 51$). Additionally, to control for participants' musical aptitude, I used scores from a music aptitude test as a covariate. I established the equality of covariance matrices by using the Box M test (Box M = 7.95, $p = .25$). I found that the overall MANCOVA was significant ($\lambda = .576$, $F = 23.197$, $p < .001$, partial $\eta^2 = .24$). In order to determine which mean differences contributed to the significant multivariate outcome, I conducted follow-up univariate ANCOVAs. I established the assumption of equal variance by using the Levene test and found that both dependent variables met the assumption (Pitch: $F = .046$, $df = 2$, $p = .96$, Rhythm: $F = 1.49$, $df = 2$, $p = .23$). I found that groups were different only regarding the outcome of the pitch test ($F = 53.39$, $df = 2$, $p < .001$, partial $\eta^2 = .42$), but not rhythm ($F = .314$, $df = 2$, $p = .73$, partial $\eta^2 = .004$). The moderate effect size suggests some level of practical significance to this finding as well. Using pairwise comparisons with a Bonferroni adjustment, I found that participants who learned by using solfège (*Adj. M* = 8.26, *SE* = .14) scored significantly higher in regard to pitch than those who learned by using letter names (*Adj. M* = 7.00, *SE* = .14) or numbers (*Adj. M* = 6.12, *SE* = .15). Those who learned by using letter names scored significantly higher on the pitch test than those who used numbers as well.

- Dependent Variable 1 "Theory": middle school students' scores on a basic music theory test (scaled 1–20).
- Dependent Variable 2 "Interest": middle school students' self-reported interest in continuing to study music in high school (scaled 1–10).
- Independent Variable "Ensemble": type of class they are in (1 = Band, 2 = Orchestra, 3 = Choir, 4 = Guitar)
- Covariate "Grade": Middle school students' current grade in music class (scaled 00–100)

Once you have computed the most appropriate statistical tests, write a report that includes any descriptive statistics necessary to understand the MANCOVA as well as the test itself and any additional information needed.

13 | Regression Analysis

Use of Regression Analysis

Researchers can use regression analyses as a means of finding relationships between variables. Many aspects of regression are similar to those of correlation analysis. However, three major differences exist. The first difference is the purpose. While correlation analyses are used to determine relationships and the strength or magnitude of relationships between variables, researchers use regression analysis as a means to predict, on the basis of information from another variable (or more), the outcome of one variable. Otherwise, researchers can use regression analysis as a means to explain why, on the basis of any given predictor variables, differences exist between participants. The second difference between regression and correlation is in what the variables are called. In a regression analysis there is an important distinction that does not exist in a correlation: there must be a dependent variable as the researcher is trying to predict, on the basis of a predictor variable, the outcome of that variable, whereas in a correlation, the relationship is not directional. The final difference between correlation and regression is the researcher's ability to infer beyond the strength of relationships through the use of such tools as regression coefficients, intercepts, and change in the regression coefficient.

As seen in Table 1.6 in Chapter 1, researchers using regression analysis are trying to answer one research question: to what extent can the one independent variable predict the outcome of the dependent variables?

For example, a music education researcher may be interested in discovering whether or not orchestra students' scores on a bowing test and a sight-reading test can predict their audition scores. In this case, the researcher would use the scores of the bowing test and sight-reading test as the independent variables, also known as predictor variables in regression analysis, as they are the variables used to "predict" the outcome of the dependent variable. In this example, the dependent variable would be the audition score, which is the score we are trying to predict by using the information from the other variables.

When considering regression analysis, researchers most commonly utilize one of two different types of regression analyses:

- Bivariate regression—similar to a bivariate correlation, as it involves two variables. In a bivariate regression analysis, one of the variables needs to be identified as the dependent variable and one as the independent variable. Keep in mind that sometimes researchers also call the dependent variables *criterion, outcome,* or *response* variable, while the independent variable is often called the *predictor* or *explanatory* variable. This type of regression analysis is most often used to examine how well one variable can predict the outcome on another variable.
- Multiple regression—used when only one dependent variable exists, but more than one independent variable is used. Three types of multiple regression analyses exist:
  - Simultaneous multiple regression—all of the data associated with the independent variables are included at the same time. This model is most often used when the researcher has little or no preconceived idea as to how the variables would impact the criterion variable, but believes them all to be related in some way.
  - Stepwise multiple regression—the independent variables are added in order of how much variance is accounted for. This model allows for the most impactful predictor variables to be used first so that the researcher can use the fewest number of variables as possible to account for the most variance as possible.
  - Hierarchical multiple regression—the researcher has greater control over the order of independent variables and includes the variables she wishes to control for first. This model is most commonly used when the researcher has some hypothesis, usually supported by previous research, that some predictor variables are more important than others, as the model allows the researcher to designate which variables she should include in which order.

Although researchers can use other types of regression analyses, such as logistic regression, which is used when the dependent variable is categorical and dichotomous, we will focus on the two most commonly employed versions: bivariate regression and multiple regression.

In Example 13.1, we can see an example of a simultaneous multiple regression. It is a multiple regression because more than one independent or predictor variable is being used to try to predict the single dependent or criterion variable.

In Example 13.1, the researchers used a multiple regression to see if different personality traits could predict injury in participants. The researchers

EXAMPLE 13.1 REPORT OF A SIMULTANEOUS MULTIPLE REGRESSION

A simultaneous multiple-regression analysis was conducted to evaluate how well the personality concurrently predicted marching music injury. The predictors were the five personality variables on the APSl, while the criterion variable was the injury index. The linear combination of the personality variables was significantly related to the injury index: $F(5,149) = 4.46$, $p < 0.001$. The multiple-correlation coefficient was $R = 0.36$, indicating that 13% of the variance of the injury index in the sample can be accounted for by the linear combination of the personality scales.

Levy, J., Kent, K. N., & Lounsbury, J. W. (2009). Big Five personality traits and marching music injuries. *Medical Problems of Performing Artists, 24*(3), 135–140.

EXAMPLE 13.2 REPORT OF A STEPWISE MULTIPLE REGRESSION

To identify the best prediction model for instrumental jazz improvisation the five independent variables that were most highly correlated with the composite improvisation scores were used in stepwise multiple regression analyses. Jazz theory achievement, aural skills, aural imitation, self-evaluation of improvisation, and improvisation class experience served as predictor variables. The results of these five stepwise regression analyses revealed the best prediction model, as that in which self-evaluation of improvisation, aural imitation, and improvisation class experience explained a total of 66% of the variance. Self-evaluation of improvisation was the strongest predictor, contributing 53% of the variance. Aural imitation contributed an additional 8%. Improvisation class experience was the third variable to enter the model, contributing 5% of the variance. Jazz theory achievement and aural skills did not enter the prediction model.

May, L. (2003). Factors and abilities influencing achievement in instrumental jazz improvisation. *Journal of Research in Music Education, 51*(3), 245–258.

found that their predictor variables were significantly related to the criterion variable and could account for 13% of the variance (i.e., could explain 13% of the cause for injury).

In Example 13.2 we can see an example of a stepwise multiple regression. This researcher employed the regression analysis to try to find the best predictors for jazz improvisation success. The researcher found that the best model included self-evaluation, aural imitation, and improvisation class experiences, which were the best predictors, in that order. Overall, the model could predict 66% of the variance. However, the most variance was explained by self-evaluation (53%), followed by aural imitation (8%), and improvisation class experience (5%). The stepwise analysis did not just create an overall model of prediction, but looked for the most important variables first in order to explain the most variance.

Regression Analysis Assumptions

Regression analysis has several assumptions. Luckily, most of them can be analyzed with relative ease either through the examination of scatterplots or through information collected via the analysis. The assumptions of regression analysis are the following:

1. Linearity—the relationship between dependent and independent variables is linear. That is, if we create a scatterplot of the two variables together, they will create a line rather than a significant curve. See Figure 13.1.
2. Homoscedasticity—the variance remains relatively equal along the entire slope of the data. See Figure 13.2.
3. Normality—as with all parametric tests, the variables have normal distributions.
4. Number of variables—it is often recommend that one should have at least 10 to 20 times as many observations (cases, respondents) as you have variables; otherwise, the estimates of the regression line are probably very unstable and unlikely to replicate if you conduct the study again.
5. Multicolinearity—no two independent variables are highly correlated (therefore redundant). If two independent variables were highly correlated, they would not really add new information to the model.

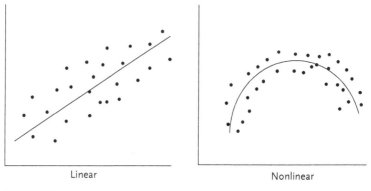

FIGURE 13.1

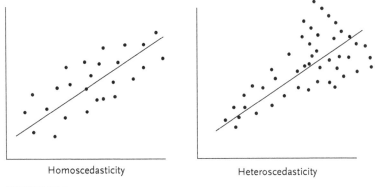

Homoscedasticity   Heteroscedasticity

FIGURE 13.2

Understanding Some of the Math Used in Regression Analysis

The formula for a basic regression model is relatively simple:

Unstandardized regression formula: $Y = a + bX$

where a is the intercept or where the regression line in the scatterplot would intersect the ordinate. Therefore, a is the value of Y when $X = 0$, where b is the regression coefficient or. Moreover, b is the *slope* of the regression line. This number indicates how many units change in the dependent variable for any one change in the independent variable.

You can compute the line of best fit or b (the slope) with a relatively simple formula:

$$b = r\left(\frac{\sigma_y}{\sigma_x}\right)$$

In this formula, the standard deviation of the y-axis values is divided by the standard deviation of the x-axis values and then multiplied by the correlation coefficient between the two scores. Keep in mind that the resulting number could be negative, which indicates an indirect relationship and therefore a slope that would move down as it moves from left to right.

In a regression analysis, we may also want to standardize the formula in order to help make analysis more interpretable. The standardized regression formula is relatively simple as well.

Standardized Regression formula: $Z_y = \beta Z_x$

This formula differs in that it uses Z scores rather than raw scores for X and Y, does not have a constant intercept, and uses β instead of b. The β is referred to as a beta weight rather than a regression coefficient.

Null Hypothesis of the Regression Analysis

The null hypothesis of a regression analysis is that the predictor variable or variables are not able to predict the dependent variable. It could be written as

$H_0: \beta_1 = 0$

This null hypothesis is directly stating that the slope or standardized beta weight is 0 and that the predictor variable had no impact on the dependent variable. The same could be stated for a multiple regression. If, for instance, the model included two predictor variables, the null hypothesis could be stated as

$H_0: \beta_1 = \beta_2 = 0$

In our musical example of trying to see if a bowing test and sight-reading test could predict audition score outcomes, our null hypothesis could be stated as

$H_0: \beta_{\text{bowing test}} = \beta_{\text{sight-reading test}} = 0$

With this statement, the researcher would be assuming that neither the bowing test nor the sight-reading test impacted the slope and therefore could not predict audition scores.

The alternative hypothesis for the multiple regression could be stated as

$H_a: \beta_1 \text{ and/or } \beta_2 \neq 0$

In this alternative hypothesis, the researcher is claiming that at least one of the beta weights corresponding to the predictor variables is able to predict, to some degree, the dependent variable as a slope is created.

In our musical example, the alternative hypothesis could be stated as

$H_a: \beta_{\text{bowing test}} \text{ and/or } \beta_{\text{sight-reading test}} \neq 0$

Finally, the directional hypothesis for the multiple regression analysis could be stated as:

$H_1: \beta_1 \text{ and/or } \beta_2 > 0$

The direction of the hypothesis could go in either direction, based on information the researcher discovered from any previous research.

In our musical example, the directional hypothesis could be stated as

$H_a: \beta_{\text{bowing test}} \text{ and/or } \beta_{\text{sight-reading test}} > 0$

Regression Analysis Research Design

Although regression analysis can be used in experimental designs, it is most commonly employed in descriptive studies in which researchers have collected some data that they believe will be able to predict another variable. For example, a music education researcher may conduct a survey and be interested in whether or not music teachers' ages are able to predict their self-reported job satisfaction. As these variables are both scaled, continuous data, it would be appropriate to use a regression analysis. A researcher can use a regression analysis in an experimental design as well. For example, a music education researcher may be interested in exploring how well different counting systems work in predicting students' ability to sight-sing music. You may be thinking that this example sounds somewhat like an ANOVA design and you would be correct. The difference is in how the logic is employed and what can be inferred fro the different analyses. In the ANOVA design, we would be able to say that students who received instruction using numbers scored higher or lower on a sight-singing test than students who learned via syllables. In a regression analysis, we would be able to say that overall, the method of learning counting through syllables was better able to predict scores on a sight-singing test than learning counting through numbers.

Understanding All of the Symbols

Before we continue, it may be beneficial to pause and reflect on the use of all of the different symbols employed when you are reporting regression analysis. Often, readers need to have only a few Greek or mathematical symbols in mind when they are reading a research report using statistical procedures. Regression analysis, however, has several layers of reporting in which researchers use symbols as a means of reporting their findings.

When reporting regression analyses, many researchers focus on the reporting of β or b. A negative β indicates that an inverse relationship exists between the two variables. The beta weight reports the magnitude of the slope between the variables. That is, a beta weight of 1 means that for every 1 unit of measurement (whatever the unit of measurement may be) the x-axis increases, the y-axis increases by 1 unit of measurement as well. Researchers also report the value of r, which is the correlation between the two variables, or r^2 (or even Adjusted r^2 in a multiple regression). The r value helps the researcher understand how well the regression model works by quantifying how well the predicted scores match with actual reported scores. The r^2 value indicates the amount of variance in the dependent variable that is explained by the independent variable. When conducting a multiple regression, researchers often use the Adjusted r^2, as it accounts for multiple independent variables and is more conservative.

Setting Up a Database for a Multiple Regression

The database for a regression analysis should have at least two columns for a bivariate analysis or at least three columns for a multiple regression. In

	SIGHTREAD	BOWINGTEST	AUDITION
1	1	3	34
2	4	3	34
3	3	3	55
4	1	2	43
5	4	2	63
6	3	3	34
7	3	3	34
8	2	3	54
9	4	3	23
10	1	4	55
11	2	3	22
12	3	2	15
13	4	2	23
14	1	3	54
15	2	4	34

FIGURE 13.3

Figure 13.3, we can see a database set up to run a multiple regression with one dependent variable (audition) and two independent variables (sight-reading test score and a bowing-test score).

Conducting a Regression Analysis in SPSS

For this section, please use the provided example dataset (Regression Analysis Example 1 ▶) to complete the analysis while we are reading. This exercise will help give you some experience in conducting a regression analysis.

In order to complete the multiple regression in SPSS, follow these steps:

1. Click **Analyze**.
2. Click **Regression**.
3. Click **Linear**.
4. Insert the dependent variable in the **Dependent** variable space and the independent variables in the **Independent(s)** variable space, as seen in Figure 13.4.
5. Click **Plots** and select all plots desired.
6. Click **Continue**.
7. Click **Statistics**.
8. Click on **Descriptives** and **Collinearity Diagnostics**.
9. Click **Continue**.
10. Click **OK**.

Once you have clicked **OK**, SPSS will create five boxes that provide all of the information you will need to understand your data and create a research report. The first box is a simple reporting of the descriptive statistics (see Figure 13.5).

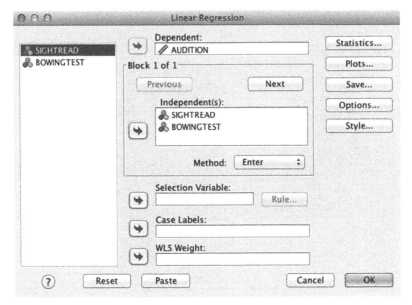

FIGURE 13.4

Descriptive Statistics

	Mean	Std. Deviation	N
AUDITION	61.63	24.975	54
SIGHT READ	3.37	1.496	54
BOWINGTEST	4.91	2.191	54

FIGURE 13.5

The second box provides the researcher the correlations among the different variables (see Figure 13.6). Remember that we are trying to avoid high correlations between independent variables.

From this box we see that the correlation between the sight-reading and bowing test is only .29. This is a relatively weak correlation; therefore, we can continue with the analysis.

The next important box tells us about the model summary and is where we get R, R^2, or Adjusted R^2 (see Figure 13.7).

From this box we can see that the predicted scores match the actual scores well ($r = .72$) and that we have explained roughly half of the variance (r^2) with this model (.52).

The next box produced in the SPSS regression analysis is the ANOVA table that reports on the significance of the R Square and Adjusted R Square tests (see Figure 13.8). This table tells the researcher if the amount of variance explained is significant or not. This information should be reported in the research report.

Correlations

		AUDITION	SIGHTREAD	BOWINGTEST
Pearson Correlation	AUDITION	1.000	.461	.663
	SIGHTREAD	.461	1.000	.293
	BOWINGTEST	.663	293	1.000
Sig. (1-tailed)	AUDITION	.	.000	.000
	SIGHTREAD	.000	.	.016
	BOWINGTEST	.000	.016	.
N	AUDITION	54	54	54
	SIGHTREAD	54	54	54
	BOWINGTEST	54	54	54

FIGURE 13.6

Model Summary

Model	R	R Square	Adjusted R Square	Std. Error of the Estimate
1	.720[a]	.518	.499	17.675

[a]Predictors: (Constant). BOWINGTEST, SIGHTREAD

FIGURE 13.7

From Figure 13.8, we can see that the F test for the amount of variance explained is significant ($F = 27.41, p < .001$).

The final important box (see Figure 13.9) gives the researcher the beta weights for the independent variables. Remember that these weights are interpreted as a slope From this box, we can see that the standardized beta for sight reading is .29, meaning that for every unit increase in the sight-reading score, audition scores rise by .29. Similarly, we see that for every unit rise in the bowing test, the audition score rises by .58. If we visualized these scores in a graph, they would look something like the one found in Figure 13.10.

From this graph, we can see that the angle of the slope of the bowing test is greater than that of the sight-reading scores, a finding that makes sense, given the standardized beta weights.

Also, from the box found in Figure 13.9, we can analyze the appropriateness of the independent (predictor) variables (whether or not we met the assumption of colinearity) by using the tolerance and variance inflation factor (VIF).

Tolerance is $1 - R^2$ for the regression of a given independent variable on all the other independents, ignoring the dependent. There will be as many tolerance coefficients as there are independent variables. The higher the intercorrelation of the independents, the more the tolerance will approach zero. As a rule of thumb, if tolerance is less than .20, a problem with multicolinearity is

ANOVA[a]

Model		Sum of Squares	df	Mean Square	F	Sig.
1	Regression	17126.156	2	8563.078	27.411	.000[b]
	Residual	15932.437	51	312.401		
	Total	33058.593	53			

[a]Dependent Variable: AUDITION
[b]Predictors: (Constant), BOWINGTEST, SIGHTREAD

FIGURE 13.8

Coefficients[a]

Model		Unstandardized Coefficients		Standardized Coefficients	t	Sig.	Collinearlty Statistics	
		B	Std. Error	Beta			Tolerance	VIF
1	(Constant)	12.873	7.197		1.789	.080		
	SIGHTREAD	4.879	1.697	.292	2.874	.006	.914	1.094
	BOWINGTEST	6.585	1.159	.578	5.683	.000	.914	1.094

[a]Dependent Variable: AUDITION

FIGURE 13.9

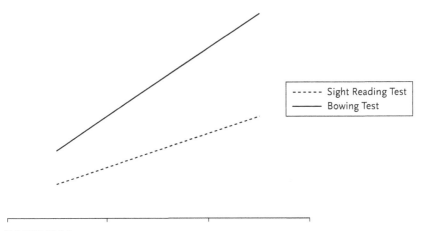

FIGURE 13.10

indicated. For this running example, no multicolinearity problem is indicated. When tolerance is close to 0 there is high multicolinearity of that variable with other independents and the b and beta coefficients will be unstable. The more the multicolinearity, the lower the tolerance, the higher the standard error of the regression coefficients.

VIF is simply the reciprocal of tolerance. Therefore, when VIF is high there is high multicolinearity and instability of the b and beta coefficients. As a rule

> **EXAMPLE 13.3 POSSIBLE REGRESSION REPORT**
>
> In order to examine the predictive relationship between a bowing and sight-reading test on subsequent audition scores, we conducted a multiple regression. Because no a priori theory existed, we employed a simultaneous multiple regression by using bowing and sight-reading test scores as the predictor variables and an audition score as the dependent variable. We found that participants ($N = 54$) observed scores and predicted scores were highly correlated ($r = .72$) and that both the sight-reading test scores ($\beta = .29$, $t = 2.87$, $p = .006$) and bowing test scores ($\beta = .58$, $t = 5.68$, $p < .001$) were significantly related to audition scores. We analyzed the appropriateness of the model and found that the predictor variables met the assumptions of regression analysis, as indicated by a tolerance of .91 and a VIF of 1.09, and found that a significant amount of variance (Adjusted $R^2 = .50$, $F = 27.41$, $p < .001$) was the result of the regression.

of thumb, VIF greater than 4.0 (not the case in this example) indicates a multicolinearity problem.

Reporting a Regression Analysis

Once you have conducted the regression analysis and interpreted the outcomes, you can create a research report that can give the readers all of the information they need to be able to understand, critique, and apply your findings. Given the dataset in our running example, one possible write-up of the data may look like the one found in Example 13.3.

Regression-Analysis Writing Hints

From the research report in Example 13.3, we see some hints that might help you state your findings as clearly as possible while providing all of the necessary information.

- Make sure that you provide the following in your report:
  - N
  - r
  - Discussion of the type of regression analysis used
  - β, t, and p for each predictor variable
  - R Square or Adjusted R Square and F
  - Multicolinear analysis including tolerance and variance inflation factors
- Per American Psychological Association guidelines, use first-person pronouns for clarity, especially when describing actions you took, while keeping all of your actions in past tense.

- Use the active voice.
- Avoid personification.
- Make sure all statistical indicators are in italics.
- Keep your actions as researcher in the past tense.

Regression-Analysis Exercises

Given the provided dataset, Regression Example 2 ⊚, compute the requisite tests. The columns included in this dataset are three predictor (independent) variables, including scores on a rhythm test, harmony test, and musical expression test. The dependent variable is a performance test.

Once you have computed the most appropriate statistical tests, write a report that includes all of the information required for a reader to assess the accuracy of your work as well as the possible implications that may exist.

14 | Data Reduction
FACTOR AND PRINCIPAL COMPONENT ANALYSIS

Use of Data Reduction Techniques

Researchers use data reduction techniques for one of two primary reasons:

1. To reduce a large number of variables into a smaller number of variables in order to facilitate subsequent statistical analysis.
2. To examine the underlying structures of a given phenomenon by classifying the variables that make up the whole.

As seen in Table 1.6 in Chapter 1, researchers using factor analysis are trying to answer one of two questions: given a large number of variables collected, what are the underlying structures or fewest number of latent variables (inferred from the data rather than overtly collected) that really exist?

For example, a researcher may be interested in the underlying occupational identity of in-service music educators. The researcher may have asked several questions in a survey about the extent to which they believed themselves to be, for instance, educators, musicians, conductors, or entertainers. If the researcher is interested in finding out, through latent variable analysis, what really is the fundamental occupational identity of the participants, he is trying to answer the question regarding the underlying structures of participants' occupational identity. In this instance, the researcher would most likely use factor analysis. If, however, the same researcher was primarily interested in reducing the data for further analysis and wanted the smallest number of uncorrelated variables from the data, he would use a principal component analysis (PCA).

Generally speaking, when researchers are trying to reduce a large number of variables into a smaller number of variables, they use a PCA. PCA seeks a linear combination of variables such that the maximum variance is extracted from the variables. That is, researchers using PCA are trying to find the most explanation as possible, based on the variables included in the model. It then removes this variance and seeks a second linear combination, which explains the maximum proportion of the remaining variance, and so on. This process is called the principal axis method and results in orthogonal (uncorrelated) factors (see Example 14.1).

> **EXAMPLE 14.1 REPORTING OF A PRINCIPAL COMPONENTS ANALYSIS**
>
> Responses to the PAI-R were then submitted to principal component analysis. Cattell's scree test, Kaiser's eigenvalue-greater-than-one rule, and Horn's parallel analysis (Horn, 1965) indicated that the data was best represented by a single underlying component. This component accounted for 45.8% of the variance in the PAI-R scores and all the pattern/structure coefficients (loadings) exceeded 50 (see Table 1).
>
> ---
>
> McCambridge, K., & Rae, G. (2004). Correlates of performance anxiety in practical music exams. *Psychology of Music, 32*(4), 432–439.

> **EXAMPLE 14.2 REPORTING OF A FACTOR ANALYSIS**
>
> Twenty-one top barrier items were also selected for an exploratory FA. The subject-to-variable ratio was 3:1. A scree plot was used to determine that a two-factor solution would be most appropriate. Upon an initial examination of communalities, six barrier items (Keeping your marriage, partnership, relationship, etc., together; Accumulation of debt due to school loans; Moving away from friends and family; Having an impact on K–12 students now; Being awarded little or no financial assistance; and Residency requirement) were removed because of low communality values ($< .300$). The remaining 15 items produced a Kaiser-Meyer-Olkin measure of sampling adequacy of .812; Bartlett's test of sphericity was found to be significant. Therefore, the sample of barrier items was determined to be appropriate for an FA procedure. Two factors were rotated by using a direct oblimin rotation method ($\delta = .3$). The category "Financial Challenges" was found to contribute 40.74% of the variance, while "Family/Time Considerations" was found to contribute 18.32% of the variance. Combined, both factors contributed 59.04% cumulative variance to the solution (see Table 4).
>
> ---
>
> Teachout, D. (2004). Incentives and barriers for potential music teacher education doctoral students *Journal of Research in Music Education, 52*(3), 234–247.

Researchers trying to find the underlying structure of a phenomenon most often use a *factor analysis* factor analysis (FA), also called principal factor analysis (PFA) or principal axis factoring (PAF), which seeks the least number of factors that can account for the common variance (correlation) of a set of variables (see Example 14.2). The distinction between these uses of factor analysis techniques will become clearer as we move through the chapter.

	factor	
Item	1	2
1. Financial Challenges		
Leaving a good K-12 level salary	.925	−.186
Reduction of income while working on the degree	.843	−0.07
Long-term salary difference between (high-paid) K-12 level and (lower-paid) college level	.828	.000
Leaving current K-12 level job security	.815	.000
Anxiety over leaving my current K-12 teaching position	.802	−.004
Lack of paid insurance and/or pension	.664	.008
Pay, associated with completing the degree, does not match the experience and education	.635	.156
Current job time demands	.510	.338
Solution variance contributed 40.72%		
2. Family/Time Considerations		
Family obligations	−.161	.922
Yours family is currently top priority	−.170	.919
"Spinning all of the plates" (being a wife/husband, mother/father, son/daughter, teacher, volunteer, etc.)	.001	.844
Moving my family to attend school	−.162	.668
Not willing to move for a job	.292	.533
Completing the coursework while working part-time or full-time	.305	.520
Needing time to research and write the dissertation	.257	.519
Solution variance contributed 18.32%		
Total solution variance contributed by 2 factors = 59.04%		

FIGURE 14.1

Note that in both examples, the authors refer to a table. Researchers need to report the factor loadings found in the FAs technique in order to allow readers the opportunity to examine the trustworthiness of the report. This is one of the drawbacks discussed below, as it relies upon a heuristic interpretation of the researcher. See Figure 14.1 for how to report the results of an FA in table form.

Data Reduction Terminology

Before we begin to examine FA more closely, it would be helpful to be more familiar with the terminology used in FA reporting.

Factor loadings—the correlation coefficients between the variables (rows) and factors (columns). Also called component loadings in PCA. Analogous to Pearson's r, the squared-factor loading is the percent of variance in that indicator variable explained by the factor. An observed variable "loads" on a factor if it is highly correlated with the factor (has a large eigenvalue).

Eigenvalue—a numeric estimation of how much variation each component explains.

When you are interpreting factor loadings, by one rule of thumb in confirmatory FA, loadings should be .7 or higher to confirm that independent variables identified a priori are represented by a particular factor, on the rationale that the .7 level corresponds to about half of the variance in the indicator being explained by the factor. However, the .7 standard is a high one and real-life data may well not meet this criterion; thus some researchers, particularly for exploratory purposes, use a lower level, such as .4 for the central factor and .25 for other factors, and call loadings above .6 "high" and those below .4 "low." In any event, factor loadings must be interpreted in the light of theory, not by arbitrary cutoff levels.

In oblique rotation, you get both a pattern matrix and a structure matrix. The structure matrix is simply the factor-loading matrix as in orthogonal rotation, representing the variance in a measured variable explained by a factor on both a unique and common contributions basis. The pattern matrix, in contrast, contains coefficients that just represent unique contributions. The more factors, the lower the pattern coefficients as a rule, since there will be more common contributions to variance explained. For oblique rotation, the researcher looks at both the structure and pattern coefficients when attributing a label to a factor.

Clarifying the Uses of Data Reduction

As we have already discussed the two primary types of data reduction techniques, we can move on two the different types of FA itself. An FA can be used in one of two ways. Researchers who are trying to examine an existing hypothesis or theory will most likely utilize a *confirmatory* FA, while those researchers examining a dataset without an a priori theory will most likely use an *exploratory* FA. The principles remain the same. In a confirmatory factor analysis, the researcher purposefully inputs a set number of variables or forces a set number of factors in order to examine an already established theory. For example, if a researcher has information from previous research that three different categories of culturally responsive teaching existed, he might force the data into three categories in order to see if the groupings of variables are logical and confirm the previously existing theory of culturally responsive teaching in the area of music.

Moreover, social science researchers also often utilize a rotational method that helps data fit into statistical models. The two types of rotations most commonly used are *orthogonal* rotations and *oblique* rotations. Orthogonal rotations limit the correlation allowed among variables, thus producing uncorrelated factors, while an oblique rotation allows for factors to be correlated to one another. When researchers do not allow a rotation to occur, most items load on the early factors, and usually have many items load substantially on more than one factor. Rotation serves to make the output more understandable, by seeking so-called simple structure, a pattern of loadings in which items load most strongly on one factor, and much more weakly on the other factors.

The most common orthogonal rotation is a varimax rotation, which rotates the factor axes to maximize the variance of the squared loadings of a factor (column) on all the variables (rows) in a factor matrix, with the effect of differentiating the original variables by extracted factor. Each factor will tend to have either large or small loadings of any particular component. A varimax solution yields results that make it as easy as possible to identify each variable with a single factor. This is the most common rotation option.

The most common oblique rotations are the oblimin rotation and the promax rotation. Oblimin rotation is the standard method when you want a non-orthogonal (oblique) solution—that is, one in which the factors are allowed to be correlated. This method will result in higher eigenvalues but diminished interpretability of the factors. Promax rotation is an alternative non-orthogonal (oblique) rotation method that is computationally faster than the direct oblimin method and therefore is sometimes used for very large datasets.

Potential Limitation of Data Reduction

There is one issue that can be seen as a limitation of FA: the heuristic nature of the interpretation of the data. In fact, the use of heuristics is a possible downfall of FA. A heuristic is a means to solve a problem without the exact information available. Trial and error is an example of a humanly created heuristic. Heuristics often rely upon experience-based learning and logic rather than data-driven logic. This aspect is most evident in the FA in that it is the responsibility of the researcher to interpret the findings of an FA. Moreover, a different researcher might interpret the exact same findings in a different manner. This limitation is an issue, mitigated, however, by ethical and full reporting of the FA outcomes and interpretations so that readers may examine how the researcher analyzed the data.

Assumptions of Data Reduction

Factor analysis is relatively free of assumptions. However, three primary assumptions of FA exist.

1. Linearity—the relationship between variables should be linear (similar to the regression analysis).
2. Factorability—tests how well the factors relate and if the analysis can be trusted. Factorability is most commonly tested by using two measures of sampling adequacy:
    a. Kaiser-Myer-Olkin (KMO)—should be greater than at least .5, but even better at above .08. The KMO test basically tells us if factors are

going to exist in the analysis and does so by examining any shared variance between different variables. If no shared variance exists, it would be unlikely that variables would be able to be grouped together to form components or factors.
    b. Bartlett's test of sphericity—should be significant, aka < .05
3. Sample size—some researchers have suggested that you need as many as 20 participants for every variable included in the factor analysis. Some have gone as low as 3:1. Most would agree, however, that 10:1 is an adequate participant to variable ratio. That is, if you want to include 10 variables in the analysis, you should have at least 100 participants.

Research Design and Null Hypothesis of Data Reduction

Factor-analysis techniques are not really associated with any given research design. They are used more generally as a means of reducing data to better understand a large amount of data or to use in subsequent analyses. If a researcher is employing FA, it is highly unlikely that any version of an experimental design would be appropriate. This type of analysis is primarily employed in descriptive studies. As such, FA does not really have a null hypothesis. One could state that the null hypothesis is that no factors exist. However, if that were true, the analysis itself would be invalid, as factorability is an assumption of the test.

Setting Up a Database for Data Reduction

Because there are no independent or dependent variables in an FA (there are only variables that a researcher would like to reduce into a smaller number of latent variables), a database needs to include only the variables that the researcher wishes to reduce or examine (see Figure 14.2). In this database we can see a series of items in which study participants indicated the extent to which they identified as different occupational identities.

Conducting a Data Reduction Technique in SPSS

For this section, please use the provided dataset, Data Reduction Example 1 ⓑ, in order to complete the analysis as you read. This exercise will give you some experience in conducting the statistical analysis prior to completing one on your own. In order to conduct a factor analysis in SPSS, you should take the first step of deciding upon what type of factor analysis you wish to conduct. Table 14.1) may help you decide.

	self educator	self musician	self perform	self ME	self artist	self musictch	self music scholar	self ensemdir	self compose	Self conductor
1	5	6	6	6	5	5	5	6	3	5
2	6	6	6	6	5	6	5	6	3	6
3	6	6	5	6	4	5	5	6	6	4
4	6	6	5	6	5	6	6	5	4	5
5	6	6	6	6	6	6	6	6	2	6
6	6	6	4	6	6	6	4	6	3	6
7	4	4	.	5	3	4	2	4	3	5
8	6	6	6	6	6	6	5	6	2	6
9	6	5	4	6	2	6	2	6	2	6
10	6	6	5	6	5	6	6	6	5	6
11	6	6	6	6	5	6	5	6	5	5
12	6	6	6	6	3	6	3	6	3	5
13	6	6	5	6	6	6	3	6	3	3
14	5	6	6	6	4	4	4	4	3	3
15	6	6	6	6	5	6	4	.	1	6
16	6	6	6	6	6	6	4	6	2	6
17	6	6	6	6	6	6	6	6	6	6
18	5	6	6	6	6	6	5	6	1	6

FIGURE 14.2

TABLE 14.1

DESIRED OUTCOME	FACTOR ANALYSIS TECHNIQUE
Reduce a large number of variables into a small number	Principal component analysis
Examine the underlying structure of a group of variables	Factor analysis
DESIRED OUTCOME	ROTATION METHOD
Allow for correlated factors or components to make sure no data is lost	Oblique: Promax or oblimin
Do not allow correlated factors so that you create disparate variables best used for subsequent analyses	Orthogonal: varimax

Once you have decided on the most appropriate method for your analysis, follow these steps to conduct a factor analysis. For the purposes of this example, we will be conducting an oblique PCA by using a promax rotation.

1. Click **Analyze**.
2. Click **Dimension Reduction**.
3. Click **Factor**.
4. Move all of the variables you wish to examine into the variables list (in this case, include all of the variables).
5. Click on **Descriptives**.
6. Make sure to select **KMO** and **Bartlett's Test for Sphericity**, then select **Continue**.
7. Click on **Extraction**.
8. Under the **Method** drop-down box, select **Principal Components Analysis** (keep in mind that the default setting for SPSS is a **Principal Components Analysis**).
9. Under **Extract** make sure that you select **Eigenvalues over 1**. If you were conducting a confirmatory factor analysis, you could force SPSS to fit the data into a specified number of factors rather than by the eigenvalues.
10. Click **Continue**.
11. Click on **Rotation**.
12. Select **Promax** (an oblique rotation allowing for correlation between factors).
13. Click **Continue**.
14. Click on **Options**.
15. Click on **Sorted by Size** (this step improves the readability of the output).

16. *Click on* **Suppress Absolute Values** *and enter the number* .30. *N.B. This step helps improve the readability of the output, and most authors agree that anything below .30 does not constitute a crossloading.*
17. *Click* **Continue**.
18. *Click* **OK**.

Once you have clicked on **OK**, there are four boxes that we will need to know about to be able to report our data. The first box is the KMO and Bartlett's test for sphericity (see Figure 14.3).

From this box, we see that the KMO measure of sampling is (rounded) .87. As we learned above, we are aiming for a minimum number above .50. So, in this example, we are doing well. The KMO measure of sampling adequacy tests whether the partial correlations among variables are small and that some factors or components exist. We can also see that we have a Bartlett's test of sphericity that is significant. This is the goal. Bartlett's test of sphericity tests whether the correlation matrix is an identity matrix, which would indicate that the factor model is inappropriate. Remember, we *want* a significant Bartlett's test. Finally, the researcher should also examine the participant-to-variable ratio, although it is not in this box. In this example, there were 300 participants and 26 included variables, in a ratio of roughly 11.5:1, which is above the suggested 10:1. On the basis of these three indicators, it is appropriate to continue with the analysis.

The next important box is the one that describes how much variance the components (factors) are able to explain (see Figure 14.4).

From this box, we can see that the analysis has identified six components that each has an eigenvalue over 1. These six components explain roughly 69% of the variance in the analysis.

The next important box to report in a PCA is the pattern matrix, which tells us which variables are loaded onto which component and the strength of the relationship between the variable and the component. Keep in mind that the factor loadings are similar to a correlation between the individual variables and the components (not between the variables themselves). See Figure 14.5.

KMO and Bartlett's Test

Kaiser-Meyer-Olkin Measure of Sampling Adequacy		.865
Bartlett's Test of Sphericity	Approx. Chi-Square	4256.253
	df	325
	Sig.	.000

FIGURE 14.3

Total Variance Explained

Component	Initial Eigenvalues			Extraction Sums of Squared Loadings			Rotation Sums of Squared Loadings[a]
	Total	% of Variance	Cumulative %	Total	% of Variance	Cumulative %	Total
1	9.179	35.303	35.303	9.179	35.303	35.303	6.734
2	3.116	11.984	47.287	3.116	11.984	47.287	6.131
3	1.797	6.913	54.201	1.797	6.913	54.201	6.071
4	1.517	5.835	60.036	1.517	5.835	60.036	3.500
5	1.307	5.027	65.063	1.307	5.027	65.063	2.708
6	1.015	3.903	68.966	1.015	3.903	68.966	3.115
7	.911	3.505	72.471				
8	.784	3.014	75.485				
9	.749	2.881	78.366				
10	.717	2.759	81.125				
11	.571	2.196	83.321				
12	.544	2.091	85.412				
13	.476	1.831	87.243				
14	.402	1.546	88.789				
15	.380	1.460	90.249				
16	.337	1.296	91.545				
17	.319	1.226	92.771				
18	.306	1.178	93.949				
19	.265	1.018	94.967				
20	.249	.959	95.926				
21	.232	.894	96.820				
22	.207	.796	97.616				
23	.194	.745	98.362				
24	.161	.618	98.980				
25	.140	.537	99.517				
26	.126	.483	100.000				

Extraction Method: Principal Component Analysis.
[a] When components are correlated, sums of squared loadings cannot be added to obtain a total variance.

FIGURE 14.4

N.B. If this were a factor analysis, you would report the rotated matrix for interpretation.

From this matrix, we can see which variables loaded onto the six different components. We can also see where some crossloadings occurred.

Pattern Matrix[a]

	Component					
	1	2	3	4	5	6
self ME	.864					
self educator	.810					
self teacher	.768					
self musictch	.751					
other musictch	.611					
other ME	.539		.464			
other educator	.531		.370		−.320	
other teacher	.490					
self conductor		.871				
selfensemdir		.801				
otherensdir		.768				
other conduct		.638				
other musician			.831			
other artist			.311			
other perform			.798		.391	
other musicscholar		.331	.564			
other compose				.878		
self compose				.870		
other entrep				.757		
self entrep				.656		
self perform					.732	
self artist					.618	
self musician	.564				.565	
self musicscholar		.422			.454	
self entertainer						.811
other entertain						.804

Extraction Method: Principal Component Analysis.
Rotation Method: Promax with Kaiser Normalization.
[a]Rotation converged in 16 iterations.

FIGURE 14.5

Crossloadings exist when one variable has relatively strong associations with more than one component. If you were using this analysis to reduce data for subsequent analysis, you might consider removing variables that have high crossloadings from any further analysis. In this example, the variable "selfmusicscholar" might be removed, as it loads relatively equally on both component 2 (.422) and component 5 (.454). Also, if you were using the FA technique to reduce data for subsequent analyses, you would need to check the internal consistency of each component or factor by using Cronbach's alpha (see Chapter 16) and you should not use any factor with a reliability of less than .60 or a component or factor that has fewer than three variables included in it. This box also informs the researcher as to how many iterations the analysis took to converge into the reported version. In this case, it took 16 rotations to find the most simple loadings.

The next step is where some of the heuristic issues exist with FA techniques. Once the researcher has decided to use all six components, he or she must decide how to describe and label the components. In this example, the researcher, on the basis of the work of other researchers, labeled the components as educator (component 1), ensemble leader (component 2), a creative businessperson (component 4), and entertainer (component 6). However, their musical identities separated into either an external music identity (component 3), in which others saw her as a musician, artist, performer, or scholar, or an internal identity (component 5), in which she saw herself differently in the same roles.

Although this interpretation is based on previous research, some other scholars may examine the pattern matrix and believe, on the basis of a conflicting or different theory or framework, that the components mean something different. This response does not have to be a negative thing and can spark interesting debate among scholars.

The final box that needs to be reported is the correlation matrix. Remember that because we utilized an oblique rotation that allows components to be correlated, we must report the component correlations. If we had conducted an orthogonal rotation, we would not need to consider or report correlations, as they would not exist (see Figure 14.6). We need to report these correlations because if two or more factors were highly correlated, it is likely that we might need to re-examine the type of analysis we used in order to obtain a more interpretable set of factors or components.

From this box we can see that the greatest correlation is .51 between components 2 and 3. This result is fine. If we had components that were

Component Correlation Matrix

Component	1	2	3	4	5	6
1	1.000	.501	.472	.019	.097	.155
2	.501	1.000	.511	.270	.145	.337
3	.472	.511	1.000	.272	.050	.420
4	.019	.270	.272	1.000	.183	.295
5	.097	.145	.050	.183	1.000	.099
6	.155	.337	.420	.295	.099	1.000

Extraction Method: Principal Component Analysis.
Rotation Method: Promax with Kaiser Normalization.

FIGURE 14.6

more highly related (roughly .70 or higher), we might have to reconsider the analysis.

Reporting a Data Reduction Technique

Once you have completed the analysis and labeled or explained the components or factors found in the analysis, you should have enough information to create a research report that will give the readers all that they need to interpret, critique, and apply your research. Given the data set in our running example, one possible write-up of the data may look like Example 14.3.

EXAMPLE 14.3 POSSIBLE PRINCIPAL COMPONENT ANALYSIS WRITE-UP

To answer the first research question, I conducted an exploratory FA of participant ($N = 300$) responses to items designed to elicit information about their perception of their own occupational identity. In this FA, I used promax rotation and KMO rather than an orthogonal rotation, which produces uncorrelated factors (Costello & Osborne, 2005). Costello and Osborne argued that "in the social sciences we generally expect some correlation among factors, since behavior is rarely partitioned into neatly packaged units that function independently of one another. Therefore using orthogonal rotation results in a loss of valuable information if the factors are correlated" (Costello & Osborne, p. 3). It is likely that participants' experiences regarding various dimensions of identity are likely to be highly correlated; therefore, I employed a promax rotation. This rotation required 16 iterations to converge. As a result of the FA (in which I used a minimum eigenvalue of 1.0), I found six factors that accounted for roughly 69% of the systematic variance in responses. As seen in Table 1 (Figure 14.7), the pattern is clear and interpretable; the majority of loadings exceeded .50 and only seven crossloadings exceeded .30. I established sampling adequacy by using the KMO measure (.87) and the assumption of sphericity by using Bartlett's test of sphericity ($\chi^2 = 4256.25$, $p = < .001$). As seen in Table 2 (Figure 14.8), despite the use of an oblique rotation, no correlation above .51 existed and the average correlation was .26.

In-service music educators saw themselves and believe others saw them as educators, ensemble leaders, creative businesspersons, and entertainers. However, their musical identities separated into either an external music identity, in which others saw them as musicians, artists, performers, or scholars, or an internal identity, in which they saw themselves differently in the same roles.

Costello, A. B., & Osborne, J. W. (2005). Best practices in exploratory factor analysis: Four recommendations for getting the most from your analysis. *Practical Assessment, Research & Evaluation, 10*(7), 1–9.

Pattern Matrix

	Component					
	1	2	3	4	5	6
self ME	.864					
self educator	.810					
self teacher	.768					
self music tch	.751					
other music tch	.611					
other ME	.539		.464			
other educator	.531		.370		−.320	
other teacher	.490					
self conductor		.871				
self ensemdir		.801				
other ensdir		.768				
other conduct		.638				
other musician			.831			
other artist			.811			
other perform			.798		.391	
other music scholar		.331	.564			
other compose				.878		
self compose				.870		
other entrep				.757		
self entrep				.656		
self perform					.732	
self artist					.618	
self musician	.564				.565	
self music scholar		.422			.454	
self entertainer						.811
other entertain						.804

Extraction Method: Principal Component Analysis.
Rotation Method: Promax with Kaiser Normalization.
ᵃRotation converged in 16 iterations.

FIGURE 14.7

Component Correlation Matrix

Co...	1	2	3	4	5	6
1	1.000	.501	.472	.019	.097	.155
2	.501	1.000	.511	.270	.145	.337
3	.472	.511	1.000	.272	.050	.420
4	.019	.270	.272	1.000	.183	.295
5	.097	.145	.050	.183	1.000	.099
6	.155	.337	.420	.295	.099	1.000

Extraction Method: Principal Component Analysis.
Rotation Method : Promax with Kaiser Normalization.

FIGURE 14.8

As you can see in Example 14.3, the researcher used two tables, Figures 14.7 and 14.8, to report the majority of data for the reader to examine. Figure 14.7 is the pattern matrix; Figure 14.8 is the correlation matrix. It is an exact replica of the one produced in SPSS.

Data Reduction Writing Hints

From the research report in Example 14.3, we see some hints that might help you state your findings as clearly as possible while providing all of the necessary information.

- Make sure that you provide the following in your report:
    - *N*
    - Discussion of the type of factor analysis used and why
    - KMO and Bartlett's test for sphericity, including χ^2 and *p*
    - Type of rotation used and why
    - Number of iterations the rotation required
    - How much variance is explained
    - Correlations between components *if* you conducted a PCA
    - The logical labels you decided to give each component
    - How you decided to keep components (e.g., eigenvalues over 1)
- Per American Psychological Association guidelines, use first-person pronouns for clarity, especially when describing actions you took, while keeping all of your actions in past tense.
- Use the active voice.
- Avoid personification.
- Make sure all statistical indicators are in italics.
- Keep your actions as researcher in the past tense.

Data Reduction Exercises

Given the second provided dataset, Data Reduction Example 2 ▶, compute the requisite tests. You are trying to answer the following research question:

What are the underlying issues regarding teacher perceptions of their jobs and environments? This research question should give a significant clue as to the best type of data reduction method to employ.

The variables are listed and described below. Remember that it will be your job to describe the components or factors that you create by using whichever FA procedure you deem most appropriate.

Report your findings both by using text and table(s) where most appropriate.

TABLE 14.2

NONMUSIC	SATISFACTION WITH NON MUSIC TEACHING DUTIES
noninstruct	Satisfaction with non instructional duties
teachload	Satisfaction with teaching load
teachassign	Satisfaction with teaching assignment
facultyinflu	Satisfaction with faculty influence level
facultyauto	Satisfaction with faculty autonomy
collabor	Satisfaction with opportunities for collaboration
stdiscipline	Satisfaction with student discipline
stmotivation	Satisfaction with student motivation
stachieve	Satisfaction with student achievement
minority	Number of minority students (percentage)
specialneed	Number of special needs students (percentage)
adminsupport	Satisfaction with administrative support
commsupport	Satisfaction with community support
relatecolleage	Satisfaction with relationships with colleagues
relateadmin	Satisfaction with relationships with administrators
relateparents	Satisfaction with relationships with parents
commit	Commitment to teaching
effective	Self efficacy of teaching
mandates	Satisfaction with mandates dictated by any agency
importfaculty	How important do other faculty believe music to be?
importstudent	How important do students believe music to be?
importparent	How important do parents believe music to be?
importadmin	How important do administrators believe music to be?
homelife	How often does you job negatively impact your home life?
satrecog	Satisfaction with the recognition you receive?
identity	Do you see yourself more as teacher or musician?
isolate	How isolated do you feel?
frustrate	How frustrated do you feel?
enjoy	How much do you enjoy teaching?
benefitsociety	To what extent does music benefit society?
benefitmusic	To what extent does music benefit music students?
benefitlearn	To what extent does music benefit learning in general?
benefitemotion	To what extent does music benefit emotional development?
benefitstudent	To what extent does music benefit students?
similarphil	How similar is your philosophy of education to your admin?
play	How important is it that students play?
read	How important is it that students read music?
analyze	How important is it that students analyze music?
evaluate	How important is it that students evaluate music?
history	How important is it that students learn about music history?
compose	How important is it that students compose?
outsidejob	How satisfied are you with your opportunities to find work outside of music teaching?
advance	How satisfied are you with your opportunities to advance within education?

The columns included in this dataset are 44 variables thought to impact the career plans of music educators and are found in Table 14.2.

Once you have computed the most appropriate statistical tests, write a report that includes all of the information required for a reader to assess the accuracy of your work as well as the possible implications that may exist.

15 | Discriminant Analysis

Use of Discriminant Analysis

Researchers use discriminant analysis, which is a multivariate statistical technique, to investigate the relationship among several independent variables and a single *categorical* dependent variable, such as membership in one of two or more groups.

For example, a music education researcher might be interested in how well the views of undergraduate music students in regard to the social importance of different roles can predict students' most desired professional role, performing musician or music educator. In this example, the researcher would use the participants' responses to how socially important different musical roles are as the independent variables and the categorical grouping of performing musician or music educator as the categorical dependent variable.

Therefore, researchers can use discriminant analysis when they are interested in seeing how well several independent factors influence a categorical variable. One of the most common uses of discriminant analysis is to examine how well different independent variables can predict which group participants will belong to (the dependent variable), as seen in Example 15.1.

Model Development in Discriminant Analysis

Researchers should not include variables in the final discriminant analysis that are not significantly associated with the dependent variable to begin with. The goal is to use only the variables that will make the most impact on the findings. For this goal, the first step in using a discriminant analysis is the model development. This first step in a discriminate analysis is to use bivariate analyses to test for the significance of a set of variables, much as with a MANOVA. If the data is of different nature, other tests of significance can be used. That is, if there are only two groups, a t-test can be used. If the data is categorical,

EXAMPLE 15.1 USE OF DISCRIMINANT ANALYSIS TO PREDICT GROUP MEMBERSHIP

To determine which variables, if any, could be used to classify Stayers from Movers and Leavers for Year 1, a predictive discriminant analysis (Huberty & Olejnik, 2006). The grouping variable (dependent variable) for this analysis was the projected career decision of the participants for the following year (stay or migrate/leave). Response variables (independent variables) included work culture, comprehensive music education philosophy, perceived importance of music education, psychological factors, student quality, praxial music education philosophy, and teacher socioeconomic background. Because of the limited number of participants who anticipated migrating or leaving the following year, a multiple discriminant analysis was not possible. Instead, participants who planned to either migrate or leave the profession were collapsed into one group to allow for statistical analysis. Four participants were excluded due to missing data. In this analysis, 275 participants were classified as Stayers and 50 were classified as Movers/Leavers.

The Box M test was not significant (Box M = 30.92, p = .414), which indicates that group variance was equivalent and the assumption of homogeneity of covariance was met. The Wilks's lambda was significant for each variable included in the model. Because this discriminant analysis had only two groups, only one function was produced. The eigenvalue for this function was .087. The Wilks's lambda test of the function (Wilks's lambda = .920, df = 7, p = .002) indicates that the single function produced by the discriminant analysis was statistically significant. The structure matrix in Table 4 shows the correlation of each variable with the single discriminant function.

This discriminant analysis model was able to classify correctly 85.5% of cases overall. The majority of Stayers were correctly classified. However, only 7.1% of Leavers and Movers were correctly classified. Table 5 shows the classification results from the Year 1 analysis.

A multiple predictive discriminant analysis was conducted to determine which variables, if any, could be used to classify participants as either Stayers, Movers, or Leavers, as projected for Year 5. The grouping variable for this analysis was the predicted career decision of the participants for 5 years in the future (stay, migrate, or leave). Response (predictor) variables included work culture, subject importance, student quality, psychological factors, professional frustrations, number of children, teaching experience, satisfaction with recognition, mentor program participation, and marital status.

The Box M test was not significant (Box M = 134.77, p = .152), indicating that group variance was equivalent and the assumption of homogeneity of covariance was met. The Wilks's lambda for each variable included in the 5-year analysis was significant, which suggests that each variable affected the overall discriminant analysis. Two different functions were produced by this discriminant analysis. The eigenvalue for Function 1 was .209, explaining 67.2% of variance. The eigenvalue for the second function was .102, explaining an additional, orthogonal 32.8% of variance. The cumulative variance explained by both functions was 100%. The Wilks's lambda was significant for both Function 1 (Wilks's lambda = .751, df = 20, p = < .001) and Function 2 (Wilks's Lambda = .908, df = 9, p = .009).

An examination of structure coefficients for each function (see Table 6) allows the researcher to assign a label to the dimension that a function measures, much like factor loadings in factor analysis. For clarity, the meta-variable corresponding to Function 1 is labeled Personal Teacher Characteristics and the meta-variable corresponding to Function 2 is labeled Professional Work Life.

The Year 5 discriminant analysis model was able to correctly classify 58.8% of cases. This model was able to classify Stayers accurately 79.7% of the time, Movers 38.7% of the time, and Leavers 35.8% of the time, findings that suggests that Stayers are more easily classified than Movers or Leavers (see Table 7).

Russell, J. A. (2008). A discriminant analysis of the factors associated with the career plans of string music educators. *Journal of Research in Music Education, 56*(3), 204–219.

Huberty, C. J., & Olejnik, S. (2006). *Applied MANOVA and discriminant analysis* (Vol. 498). John Wiley & Sons.

EXAMPLE 15.2 BUILDING A DISCRIMINANT ANALYSIS MODEL

In a continuation of the analysis described in Example 15.1, bivariate associations between each plausible variable and the career decisions of participants were then examined. Chi-square tests for categorical variables and t-tests for continuous variables were conducted to determine if there were significant differences in predictor variable scores by membership in each projected career group. Variables lacking significant differences were excluded from the final models (see Table 2). As a final model development step, only those variables with at least 280 cases were included in either model to ensure an adequate variable-to-response ratio.

Russell, J. A. (2008). A discriminant analysis of the factors associated with the career plans of string music educators. *Journal of Research in Music Education, 56*(3), 204–219.

---

a chi-square test may be used. If the data is not categorical and has more than two groupings, the most appropriate F test may be used (i.e., ANOVA, MANOVA). And yes, different types of tests can be used in the preliminary steps of the same analysis (see Example 15.2). If the MANOVA reveals statistical significance, the researcher can advance to the next step.

The Assumptions of Discriminant Analysis

There are five basic assumptions inherent in discriminant analysis.

1. Although unequal sample sizes are acceptable, sample sizes do need to be at least 20 per group. The maximum number of independent variables should be at least one quarter of the number of observations.
2. The data will represent a normal distribution. Violations of this assumption are not fatal to discriminant analysis as long as the issue is with skewness and not outliers.
3. Discriminant analysis assumes homogeneity of variances and covariances. Homogeneity of variance assumes that the amount of variability in each group is relatively equal.
4. No outliers exist in each group or variable. If outliers exist in a group, Type I error may occur, as the test is based on average variances across all groups.
5. The final assumption, non-multicolinearity, assumes that no two independent variables are highly correlated.

Meeting the Assumptions of Discriminant Analysis

The researcher can undertake several steps to ensure that he or she meets the assumptions of discriminant analysis. Researchers can recode certain variables to less exacting levels to ensure an adequate sample size for each group. The researcher can also exclude from the analyses those independent variables that are due to a lack of normality. The researcher can test for homogeneity for each variable and homoscedasticity by using a Box M test during the discriminant analyses. The researcher can also recode or remove all outliers that were recoded to create viable variables. Finally, the researcher can remove any variable found to be highly correlated with any other variable from the analysis. See Example 15.3 for a research report of meeting the assumptions of a discriminant analysis.

Research Design and Null Hypothesis of Discriminant Analysis

Discriminant analysis is not really associated with any given research design. It is used more generally as a means of examining how well a set of variables can accurately predict membership in some group. As such, it is primarily used in descriptive studies rather than any experimental design. As with factor analysis, discriminant analysis does not really have a null hypothesis, as any variable found to not be associated with the group membership would be eliminated from the analysis during the model building stage. One could state, however, that the null hypothesis is that no variables can predict group membership.

> **EXAMPLE 15.3 MEETING THE ASSUMPTIONS**
> **OF DISCRIMINANT ANALYSIS**
>
> To continue the work in Examples 15.1 and 15.2 and to meet the assumption of normality needed to run discriminant analyses, the researchers included only variables with a skewness index of less than the absolute value 1 or those whose standard error of skewness was less than three times the skewness statistic (Huberty & Olejnik, 2006). On the basis of these criteria, several variables were excluded (e.g., average class size, minority student population, teacher efficacy, and certification means).
>
> Variables included in the final models for analysis were also considered clinically plausible predictors of string music educators' career decisions. This criterion for variable inclusion has been employed by other researchers who used discriminant analyses (Bae, Zhang, Rivers, & Singh, 2007). Bivariate associations between each plausible variable and the career decisions of participants were then examined. Chi-square tests for categorical variables and t-tests for continuous variables were conducted to determine if there were significant differences in predictor variable scores by membership in each projected career group. Variables lacking significant differences were excluded from the final models (see Table 2). As a final model development step, only those variables with at least 280 cases were included in either model to ensure an adequate variable-to-response ratio.
>
> Although age was a significant factor, it was excluded from the discriminant analysis becasue of its high correlation (.82) with teaching experience. Participants with alternative certification were significantly more likely to project leaving the profession in 5 years than other participants ($F = 3.032$, $p = .050$); however, lack of normality (skewness = 1.66, SE of skewness = .143) did not allow this variable to be included in the final model.
>
> ---
>
> Russell, J. A. (2008). A discriminant analysis of the factors associated with the career plans of string music educators. *Journal of Research in Music Education, 56*(3), 204–219.
> Bae, S., Zhang, H., Rivers, P. A., & Singh, K. P. (2007). Managing and analysing a large health-care system database for predicting in-hospital mortality among acute myocardial infarction patients. *Health Services Management Research, 20*(1), 1–8.

Setting Up a Database for a Discriminant Analysis

The database for a discriminant analysis should contain one dependent variable that is categorical as well as any number of independent variables that meet the assumptions of the discriminant analysis. These independent variables should not only meet the assumptions of the discriminant analysis, but also already be found to be significantly related, as well as a logical grouping variable (see Figure 15.1).

In this database, we see that we have eight possible independent variables that are participants' responses as to the social importance of several different

	elementteach	popmusician	conductfac	classicmusician	hsmusicteach	appliedfac	msmusicteach	musicedfac	desiredrole
1	5	2	3	5	5	5	5	5	1
2	5	3	3	4	4	4	4	5	1
3	4	3	4	5	4	5	4	5	1
4	5	4	4	4	5	4	5	5	2
5	5	5	5	5	5	5	5	5	2
6	5	3	3	4	4	4	5	4	1
7	4	4	2	4	4	3	4	3	2
8	4	4	3	4	3	4	3	5	1
9	4	4	2	4	4	3	0	5	1
10	5	4	0	4	5	5	4	4	1
11	5	3	3	3	4	4	4	3	1
12	5	3	2	4	5	4	5	5	2
13	3	3	4	4	4	4	3	4	1
14	5	5	5	5	5	5	5	5	2
15	5	2	3	4	4	3	4	4	1

FIGURE 15.1

musical roles (scale of 1–5) as well as a single categorical dependent variable of whether or not these participants want to become a music teacher (1) or a performing musician (2) upon graduating from college.

Conducting a Discriminant Analysis in SPSS

For this section, please use the provided example database, Discriminant Analysis Example 1 ▶, in order to complete the analysis as you read. As stated above, the first thing a researcher wants to do is to see if the independent variables are related to the dependent variable. In this case, as there are only two different groups (those who want to become a music teacher and those who wish to become a music performer), independent sample t-tests would be the most appropriate procedure for examining group difference. (See Chapter 6 for how to run the t-tests.) Once you have completed the t-tests, you can see that only four of the eight variables differentiate between the two groups (pop musician, classical musician, high school music teacher, middle school music teacher). We will now use these categories as the independent variables in the discriminant analysis to see how well they can classify or predict group membership. In order to conduct the discriminant analysis in SPSS, follow these steps:

1. Click **Analyze**.
2. Click **Classify**.
3. Click **Discriminant**.
4. *Insert the dependent variable in the* **Grouping Variable** *slot*
5. *Insert the independent variables you wish to include in the* **Independents** *slot (see Figure 15.2).*
6. *You will need to click* **Define Range**. *In this example, those who wanted to become teachers were listed as 1 and those who wanted to become music performers were listed as 2. Therefore, the minimum is* **1** *and the maximum is* **2**. *If there were, for example, four groups labeled 1, 2, 3, 4, the minimum would still be* **1**, *but the maximum would be listed as* **4**.
7. *Once you have defined the ranges, click* **Statistics**. *You want to make sure to highlight* **Box M Test**.

8. Click **Continue**.
9. Click **Classify**.
10. Make sure that **Summary table** *is highlighted; the rest should be able to be left at the default settings.*
11. Click **Continue**.
12. Click **OK**.

Once you have clicked **OK**, you will get several boxes of information. We will, however, focus on only a few. The first piece of information we need to examine is the Box M test (see Figure 15.3). In this instance, a not significant finding indicates that group variance is equivalent and the assumption of homogeneity of covariance was met. Therefore, no significance is a good thing.

From this box, we can see that the significance is greater than .05; therefore the Box M tests was not significant, meaning that we have met the assumption of homogeneity of covariance.

The next box with useful information is the box with the eigenvalue (see Figure 15.4). Because this discriminant analysis had only two groups, only one

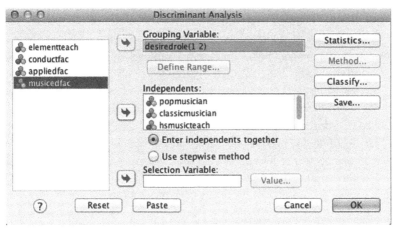

FIGURE 15.2

Test Results

Box's M		16.909
F	Approx.	1.669
	df1	10
	df2	507908.146
	Sig.	.082

Tests null hypothesis of equal population covariance matrices.

FIGURE 15.3

Eigenvalues

Function	Eigenvalue	% of Variance	Cumulative %	Canonical Correlation
1	.167[a]	100.0	100.0	.378

[a] First 1 canonical discriminant functions were used in the analysis.

FIGURE 15.4

Wilk's Lambda

Test of Function (s)	Wilks' Lambda	Chi-square	df	Sig.
1	.857	50.396	4	.000

FIGURE 15.5

function was produced. The eigenvalue for this function was .17. The larger the eigenvalue (up to 1.0), the more variance that is explained. Therefore, in this analysis we are able to explain about 17% of the difference between the groups with our analysis.

The next box with pertinent information is the Wilks's lambda information (see Figure 15.5). The larger the Wilks's lambda score, the more important the variable was to the discriminant function.

From this box, we see that the Wilks's lambda is high (.857) and that the single function produced by the discriminant analysis is able to classify groups significantly ($p < .001$).

The next important boxes are the standardized canonical discriminant function coefficient and the structure matrix (see Figure 15.6). The standardized discriminant function coefficients indicate the relative importance (largest absolute values) of the independent variables in predicting projected career decisions. The closer the function value is to an absolute value of 1, the more important a variable is in the overall function. The structure matrix shows the correlation of each variable with the single discriminant function.

From these boxes, we can see that the most important predictor for groups was the perceived social importance of the popular musician. Ironically, the least able to predict group membership was the classical-musician role. Similarly, we can see from the structure matrix that the most related to the overall function was the popular musician, while the others were all relatively related to the overall function.

The last important box for us to consider is the actual classification results (see Figure 15.7). In this box we learn how accurately the model was able to predict group membership both overall and within each group.

From this box, we see that the discriminant analysis model was able to correctly classify 72% of all the participants. This finding is well above chance. In this example, as there are only two groups, chance would be 50%. We can

Standardized Canonical Discriminant Function Coefficients

	Function
	1
pop musician	.952
classic musician	−.006
hs music teach	.008
ms music teach	.169

Structure Matrix

	Function
	1
pop musician	.986
ms music teach	.364
hs music teach	.357
classic musician	.339

FIGURE 15.6

Classification Results[a]

		desired role	Predicted Group Membership		Total
			1	2	
Original	Count	1	111	47	158
		2	46	127	173
	%	1	70.3	29.7	100.0
		2	26.6	73.4	100.0

[a] 71.9% of original grouped cases correctly classified.

FIGURE 15.7

also see that 70.3% of those who wanted to become teachers (1) were correctly classified, while 73.4% of those who wanted to become performers were correctly classified (2). Both of these findings are well above chance.

Reporting a Discriminant Analysis

Once we have completed all of the steps in conducting a discriminant analysis, including model development and the analysis itself, we will have all of the information we need to report the outcomes of the analysis so that others may assess the quality and meaning of the work. Given the data in our running example, a possible write-up of this information can be seen in Example 15.4.

As mentioned above, the correlations of each variable to the overall function could be reported in table format rather than in text. That table would most likely look the one found in Table 15.1.

> **EXAMPLE 15.4 POSSIBLE DISCRIMINANT ANALYSIS RESEARCH REPORT**
>
> To determine which independent variables, if any, could be used to classify those who wished to become music teachers from those who wished to become performing musicians, I conducted a predictive discriminant analysis. I used the projected career decision of the participants ($N = 335$) after they graduated from college or university as the grouping variable (dependent). Response variables (independent variables) included participants' perceptions of the social importance of different musical roles (i.e., popular musician, classical musician, middle school music teacher, and high school music teacher). After a priori bivariate tests, I found that the other musical roles (i.e., music education faculty, conducting faculty, elementary music teachers, and applied faculty) were not significantly related to participants' career plans and were therefore not included in the discriminant analysis.
>
> The Box M test was not significant (Box M = 16.91, $p = .08$), which indicates that group variance was equivalent and the assumption of homogeneity of covariance was met. The eigenvalue for the sole function was .17. The Wilks's lambda test of the function (Wilks's lambda = .857, $df = 4$, $p < .001$) indicates that the single function produced by the discriminant analysis was statistically significant. The structure matrix shows the correlation of each variable with the single discriminant function.
>
> This discriminant analysis model was able to classify correctly 71.9% of cases overall. The majority of those who want to become music teachers were correctly classified (70.3%), as were those who wanted to become music performers (73.4%).

TABLE 15.1

VARIABLE	FUNCTION I
Popular Musician	.986
Middle School Music Teacher	.364
High School Music Teacher	.357
Classical Musician	.339

Discriminant Analysis Writing Hints

From the research report in Example 15.4 (and Table 15.1), we see some hints that might help you state your findings as clearly as possible while providing all of the necessary information.

- Make sure that you provide the following in your report:
  - N
  - Description of the variables included and the model development process

- Description of the variables not included and why
  - Box M test and all related statistics
  - Eigenvalue (remember that this data is how much variance is explained)
  - Overall significance of the function (or functions if more than one exists) via Wilks's lambda
  - At least the structure matrix if not also the standardized canonical discriminant function coefficient (in *table* format)
  - How well the model was able to classify participants overall
  - How well the model was able to classify each individual group
  - If the analysis was able to classify above the level of chance (as determined by the number of groups)
- Per American Psychological Association guidelines, use first-person pronouns for clarity, especially when describing actions you took, while keeping all of your actions in past tense.
- Use the active voice.
- Avoid personification.
- Make sure all statistical indicators are in italics.
- Keep your actions as researcher in the past tense.

Discriminant Analysis Exercises

Given the second provided dataset, Discriminant Analysis Example 2 ▶, compute the all of the requisite tests. Remember to build the model before completing the discriminant analysis so that only significantly associated variables are included. You are trying to answer the following research question:

Can the types/topics of musical questions asked by ensemble directors help classify what type of ensemble they are leading?

The data includes a single dependent variable that is program type: 1 = Band, 2 = Choir, 3 = Orchestra. The independent variables are a list of the students' perceptions of how often the directors ask specific questions about the following:

- Tone
- Intonation
- Balance
- Instrumental/vocal technique
- Dynamics
- Style
- Articulation
- Tempo
- Precision
- Phrasing
- Interpretation

Complete the necessary and most appropriate bivariate analyses as you build your discriminant analysis model and then complete the discriminant analysis and subsequent write-up. It is likely that you will have more than one function (unlike the example above). That is fine; just report the Wilks's lambda for both and then create a heuristic label for each as you describe them, much like when you are completing a factor analysis (see Chapter 14). Remember to include all of the information required for a reader to assess the accuracy of your work as well as the possible implications that may exist. It is also likely that you will use a table to report part of your analysis.

III | Reliability Analysis

16 | Cronbach's Alpha

Use of Reliability Analysis

Researchers speak a great deal about validity and reliability in research. It can be somewhat difficult to measure the validity of an instrument. It is relatively simple, however, to objectively examine the *consistency* (aka reliability) of an instrument or scores by using objective analyses. In this section of the text, we discuss the most commonly used reliability procedures, each with a different purpose:

- Cronbach's alpha (α)—used to determine the internal consistency of a scale or subscale and is computed by using correlations among all of the items included in the scale. This method is not appropriate for determining inter-rater reliability or test-retest consistency.
- Split-half reliability—used to test the reliability between two equivalent halves of an instrument or parallel forms of an instrument. The most common version of this test is the Spearman-Brown coefficient.
- Guttman split-half reliability coefficient—this statistic is used in place of the Spearman-Brown coefficient if the parallel forms of the instrument do not have equal variances.
- Inter-rater reliability—if the variables being examined are continuous, a researcher can use a simple Pearson product-moment correlation to examine the inter-rater reliability. If the data is categorical, the most appropriate test of inter-rater reliability is kappa (κ).

Although no specific rule exists as to the requisite level of reliability, many researchers agree that an acceptable reliability is one that is above .70 (all of the discussed measures of reliability use the same scale). Researchers who are conducting exploratory research or research examining complex social or psychological phenomena rather than more overtly observable behavior, however, may deem lower reliabilities acceptable.

Assumptions of Reliability Analyses

It is rare in music education research that authors overtly report the assumptions that ground reliability analyses. However, these tests have several key assumptions.

1. Additivity—each item should have a linear relationship to the entire score and is tested via Tukey's test of nonadditivity (more to come).
2. Independence of observations—the observation of each individual should be independent of the observation of any other individual included in the analysis.
3. Uncorrelated error—the errors computed for each item should not be correlated with one another.
4. In split-half tests, individual participants must be randomly assigned to the two groups.
5. In split-half tests, the two forms of the instrument must be equivalent.
6. In split-half tests, the two forms must have equal variances.

In addition to these assumptions, the same assumptions as the Pearson product-moment correlation (Chapter 3) hold true as well:

1. The variables need to be continuous. Sometimes, researchers create "dummy" variables (i.e., group membership = 1, 2, 3, or 4) for categorical data to fit into a correlation analysis. Although this method can achieve a goal, it is more appropriate to use a categorical variable as an independent variable in an analysis of variance.
2. As with all parametric tests, the variables should be normally distributed.
3. Outliers (those data points that look nothing like the majority of data points) should be kept to a minimum. It is important to remove outliers (assuming that doing so does not negatively impact the other assumptions), as outliers can greatly influence the line of best fit, which can impact your findings.
4. The relationship between the two variables should be linear. That is, the relationship between two variables should remain constant rather. than be nonlinear. This relationship can usually be most easily seen by using simple graphic representations of the data (see Chapter 3, Figure 3.3).

Cronbach's Alpha

As discussed above, Cronbach's alpha can be used to examine the internal consistency of a scale or subscale (see Example 16.1).

As seen in Example 16.1, music education researchers often use Cronbach's alpha to help validate a scale or subscale of a test or research project. As seen in Chapter 14, researchers also use Cronbach's alpha to examine the reliability of factors or components created in factor analyses.

EXAMPLE 16.1 USE OF CRONBACH'S ALPHA TO DISCUSS SCALE RELIABILITY

Scale and subscale reliability was estimated by using Cronbach's alpha. The negatively worded items were recoded to reflect a positive wording, and the 25 item scores were summed, yielding a possible range of scores from 25 to 225. The total score demonstrated $\alpha = .92$. Five of the seven subscales showed $\alpha = .81-.92$, while Factors 5 and 6 showed a = .57 and .43, respectively.

Smith, B., & Barnes, G. V. (2007). Development and validation of an orchestra performance rating scale. *Journal of Research in Music Education, 55*(3), 268–280.

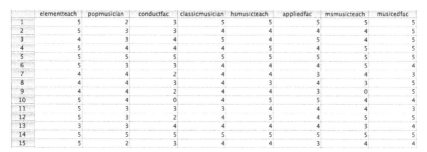

FIGURE 16.1

Setting Up a Database to Compute Cronbach's Alpha

A database that is set up to compute Cronbach's alpha should have multiple columns of data that the researcher would like to see hold together with high internal consistency. In Figure 16.1, the database consists of eight columns that contain participant responses regarding the social importance of different musical roles. The researcher is interested to see if the instrument used has internal consistency and therefore needs to compute Cronbach's alpha.

Conducting Cronbach's Alpha in SPSS

For this section, please use the provided dataset, Cronbach's Alpha Example 1 ▶, so that you can complete the analysis as you read. In order to compute Cronbach's Alpha in SPSS, follow these steps:

1. Click **Scale**.
2. Click **Reliability Analysis**.
3. Move each item included in the scale to the **Item** box.
4. Make sure to select **Alpha** from the **Model** drop-down menu.
5. Click **Statistics**.

6. *Make sure to select* **Scale if item deleted** *under* **Descriptives for** *on the left of the screen and* **Tukey's test of additivity** *on the lower right of the screen (see Figure 16.2).*
7. *Click* **Continue**.
8. *Click* **OK**.

Once you have clicked on **OK**, SPSS will produce four boxes. The first box (See Figure 16.3) is just a summary of the items included or excluded in the analysis. Items that do not have a response for each scale item will not be included in the analysis.

From this box, we can see that of the 118 participants, five participants were not included in the analysis because of some missing data.

The next box that SPSS creates is the actual value of alpha (see Figure 16.4).

From this box, we can see that eight items were included in the scale and that we have a rather high internal consistency (.79).

FIGURE 16.2

Case Processing Summary

		N	%
Cases	Valid	113	95.8
	Excluded[a]	5	4.2
	Total	118	100.0

[a] Listwise deletion based on all variables in the procedure.

FIGURE 16.3

Reliability Statistics

Cronbach's Alpha	N of items
.788	8

FIGURE 16.4

Item-Total Statistics

	Scale Mean if item Deleted	Scale Variance if Item Deleted	Corrected Item-Total Correlation	Cronbach's Alpha if Item Deleted
elementt each	28.69	14.912	.354	.788
pop musician	29.54	16.197	.136	.827
conduct fac	29.09	13.706	.586	.749
classic musician	29.15	14.093	.527	.759
hs music teach	28.62	14.149	.672	.741
applied fac	28.78	13.531	.669	.736
ms music teach	28.88	13.263	.667	.735
music edfac	28.62	15.077	.481	.767

FIGURE 16.5

What if we wanted to improve the internal consistency of a scale? The easiest way of accomplishing this step is to remove items that may not fit the scale as well as we originally believed. In order to do this, we must look at the third box created in SPSS in this analysis (see Figure 16.5).

From the right-most column in this table, we can see that alpha could be greatly enhanced if we remove the popmusician item from the scale (the alpha would increase to .83). This finding tells the researcher that the item that fits this particular scale the least is the popmusican item. We as the researchers have to decide if we wish to keep the item or remove it in order to improve alpha. In this case, as the researcher found a high internal consistency to begin with, there may be relatively little to gain from removing the item from the scale.

The final box that SPSS creates helps you analyze the assumption of additivity (see Figure 16.6). As with many of the assumptions discussed in this text (Levene, Mauchley, etc.), a significant finding indicates a violation of the assumption of additivity.

From this box (we are looking in the left-most column at the row labeled "Nonadditivity") we can see that we have a significant outcome, meaning we may have not met the assumption of additivity for this reliability analysis. If this is the case, and our final analysis is based on the reliability outcome, we may need to transform the data (in this case raise by a power of .272, as indicated) or remove more data to see if additivity can be achieved. A researcher can do so through examining scatterplots to see which variable(s) may be impacting the outcome. See Example 16.2 for a model of how to report additivity.

ANOVA with Tukey's Test for Nonadditivity

			Sum of Squares	df	Mean Square	F	sig
Between People			254.960	112	2.276		
Within People	Between Items		81.388	7	11.627	24.052	.000
	Residual	Nonadditivity	.712a	1	.712	1.474	.225
		Balance	378.274	783	.483		
		Total	378.987	784	.483		
	Total		460.375	791	.582		
Total			715.335	903	.792		

Grand Mean = 4.13
aTukey's estimate of power to which observations must be raised to achieve additivity = .272

FIGURE 16.6

EXAMPLE 16.2 REPORTING ADDITIVITY

Tukey's procedure for nonadditivity showed that an additive model was achieved for anxiety scores, $F(1, 11) = .04$, p = no significance.

Teague, A., Hahna, N. D., & McKinney, C. H. (2006). Group music therapy with women who have experienced intimate partner violence. *Music Therapy Perspectives, 24*(2), 80–86.

EXAMPLE 16.3 POSSIBLE CRONBACH'S ALPHA REPORT

Participants ($N = 118$) completed a researcher-created instrument designed to elicit their beliefs about the social importance of various musical roles. In order to examine the internal consistency of the Musical Role Social Importance Scale, I computed Cronbach's alpha. Prior to computing Cronbach's alpha, I examined the additivity of the model by using Tukey's procedure for nonadditivity and found that an additive model was achieved ($F = 1.47, p = .23$). I found a high internal consistency ($\alpha = .79$) and deemed this level of reliability appropriate for subsequent analysis.

Reporting Cronbach's Alpha

Once you have conducted all of the necessary steps and analyzed the data, you can create a research report to provide readers with all the necessary information. Given the dataset in our running example, one possible write-up of the data may look like Example 16.3.

Cronbach's Alpha Writing Hints

From the research report in Example 16.3, we see some hints that might help you state your findings as clearly as possible while providing all of the necessary information.

- Make sure that you provide the following in your report:
    - N.
    - Tukey's test of additivity, including F and p
    - Discussion of the appropriateness of subsequently using the scale or subscale.
    - The strength of the internal consistency (is it high, moderate, or weak?).
- Per American Psychological Association guidelines, use first-person pronouns for clarity, especially when describing actions you took, while keeping all of your actions in past tense.
- Use the active voice.
- Avoid personification.
- Make sure all statistical indicators are in italics.
- Keep your actions as researcher in the past tense.

Cronbach's Alpha Exercises

Given the second provided dataset, Cronbach Alpha Example 2 ▶, compute the requisite Tukey's test of additivity as well as the alpha and alpha if item deleted. The columns included in this dataset are a scale of 12 music-teacher career-commitment items (all on a scale of 1–6):

- I am Confident that I am a good musician.
- I belong in the music profession.
- I worry about being a musician.
- I am destined to be a musician.
- I am confused about being a music educator.
- Others assure me I have what it takes.
- I am comfortable with the idea of a music career,
- I am committed to a career in music.
- Others have tried to talk me out of a career in music.
- I am not comfortable with a career in music education.
- I am certain a music career will be rewarding.
- I am stressed about a music career.

Once you have computed the most appropriate statistical tests, write a report that includes all of the information required for a reader to assess the accuracy of your work, as well as the possible implications that may exist.

17 | Split-Half Reliability

Use of Split-Half Reliability

The split-half reliability method is used to test the reliability between two equivalent halves of a research instrument or parallel forms of an instrument. The split-half tests do exactly what they say they do; they take an instrument and split it in half to see if the two halves are consistent. It will randomly take half of the variables included in the analysis and compare it to the other half included in the analysis. The most common version of this test is the Spearman-Brown coefficient, as seen in Example 17.1.

Assumptions of Reliability Analyses

As discussed in Chapter 6 in regard to Cronbach's alpha, split-half reliability analysis has several assumptions:

1. Additivity—each item should have a linear relationship to the entire score and is tested via Tukey's test of nonadditivity (more to come).
2. Independence of observations—the observation of each individual should be independent of the observation of any other individual included in the analysis.
3. Uncorrelated error—the errors computed for each item should not be correlated with one another.
4. In split-half tests, individual participants must be randomly assigned to the two groups.
5. In split-half tests, the two forms of the instrument must be equivalent.
6. In split-half tests, the two forms must have equal variances.

In addition to these assumptions, the same assumptions as the Pearson product-moment correlation (see Chapter 3) hold true as well:

1. The variables need to be continuous. Sometimes, researchers create "dummy" variables (i.e., group membership = 1, 2, 3, or 4) for categorical

> **EXAMPLE 17.1 REPORT OF SPLIT-HALF RELIABILITY ANALYSIS**
>
> Effective use of this or any other measurement scale for music therapists would seem to require some form of "validation" with college-aged music therapy students in order to investigate how music therapists may be similar or different from other college-aged students and also from other college music populations. These comparisons may inform potential use of this scale as well as providing "normative" data for music therapists. The study was accomplished in two phases. The first phase of this study was to assess the scale by using different music therapy populations. Four different populations were used, representing four different geographical regions of the United States. The analysis of this data indicated that the instrument was psychometrically sound (split-half reliability ranging from .75 to .86; test-retest reliability over a four-week period was $r = .817$, $N = 58$).
>
> ---
>
> Madsen, C., & Goins, W. E. (2002). Internal versus external locus of control: An analysis of music populations. *Journal of Music Therapy, 39*(4), 265–273.

data to fit into a correlation analysis. Although this method can achieve a goal, it is more appropriate to use a categorical variable as an independent variable in an analysis of variance of some type.

2. As with all parametric tests, the variables should be normally distributed.
3. Outliers (those data points that look nothing like the majority of data points) should be kept to a minimum. It is important to remove outliers (assuming that doing so does not negatively impact the other assumptions), as outliers can greatly influence the line of best fit, which can impact your findings.
4. The relationship between the two variables should be linear. That is, the relationship between two variables should remain constant rather than be nonlinear. This relationship can usually be most easily seen by using simple graphic representations of the data.

Meeting the Assumptions of a Split-Half Reliability Analysis

Prior to computing a split-half analysis, the researcher should ensure that he has met some of the more important assumptions of the test. The first is that the split halves are equivalent forms. This assumption is tested by using Hotelling's T^2. Hotelling's T^2 is a test for equality of means between the two groups. Like previous tests of parametric assumptions, (e.g., Levene, Mauchley, Tukey's test for additivity) a significant finding means that the assumption has not been met; the two halves are significantly different. The second assumption we need to meet is that of equal variances and is tested by using the chi-square test of parallel models. Again, as with other tests of assumptions (e.g., Levene, Mauchley, Tukey's test for additivity, Hotelling's T^2), a significance of <.05 means that the assumption has not been met. Although these statistics are rarely reported in music education research, they should be included in a research report in order to give the reader full disclosure about the appropriate use of the split-half reliability analysis and results.

Setting Up a Database for a Split-Half Reliability Analysis

The database for conducting a split-half reliability will look just like one that was established for computing Cronbach's alpha. There will need to be at least two columns to compare. In Figure 17.1, we see a set of 13 variables in which participants responded to items designed to elicit information regarding how they saw themselves in different professional musical roles. The researcher is interested in how reliable the instrument was and so has decided to compute a split-half reliability analysis.

Conducting a Split-Half Reliability Analysis in SPSS

For this section, please use the provided dataset, Split Half Example 1 ⓑ, to conduct the analysis as you read. The first step in conducting a split-half reliability analysis is to examine the equal variance of the split halves by using the chi-square test of parallel models. In order to do this, follow these steps:

1. *Click* **Analyze**.
2. *Click* **Scale**.
3. *Click* **Reliability Analysis**.
4. *Move all of the variables you would like to include into the* **Items** *box.*
5. *Select* **Parallel** *in the* **Model** *drop-down menu.*
6. *Click* **OK**.

SPSS will produce three boxes. At this point, however, we are interested only in the second box, the test for model goodness of fit. See Figure 17.2.

In this box, we are most interested in the chi-square test of parallel models, as discussed above. In this example, we can see the value of the chi-square test

	selfeducator	selfmusician	selfperform	selfME	selfartist	selfmusictch	selfmusicscho...	selfensemdir	selfcompose	selfconductor	selfentertainer	selfentrep	selfteacher
1	5	6	6	6	5	5	5	6	3	5	5	4	5
2	6	6	6	6	5	6	5	6	3	6	1	4	6
3	6	6	5	6	4	5	5	6	6	4	4	4	5
4	6	6	5	6	5	6	6	5	4	5	5	4	6
5	6	6	6	6	6	6	6	6	2	6	4	3	6
6	6	6	4	6	6	6	4	6	3	6	1	5	6
7	4	4	.	5	3	4	2	4	3	5	5	5	4
8	6	6	6	6	6	6	5	6	2	6	4	4	6
9	6	5	4	6	2	6	2	6	2	6	2	2	4
10	6	6	5	6	5	6	6	6	5	6	4	3	6
11	6	6	6	6	5	6	5	6	5	5	5	5	6
12	6	6	6	6	3	6	3	6	3	5	2	2	6
13	6	6	5	6	6	6	3	4	3	3	1	1	6
14	5	6	6	6	4	4	4	4	3	3	4	3	6
15	6	6	6	6	5	6	4	.	1	6	1	1	6

FIGURE 17.1

Test for Model Goodness of Fit

Chi-square	Value	1546.334
	df	89
	Sig	.000
Log of Determinant of	Unconstrained Matrix	−8.580
	Constrained Matrix	−2.965

Under the parallel model assumption

FIGURE 17.2

is 1546.33 with 89 degrees of freedom and a significance of <.05. Therefore, we cannot assume that we have equal variance. We do not meet the assumption of equal variances. Therefore, we cannot employ the Spearman-Brown coefficient, as it assumes equal variance. We do, however, have the option of employing the Guttman split-half reliability coefficient, which is robust against the violation of equal variance.

In order to continue with the split-half analysis, follow these steps, which are the exact same even if you had found equal variance, as both Spearman-Brown and Guttman are reported in the same manner.

1. Click **Analyze**.
2. Click **Scale**.
3. Click **Reliability Analysis Statistics**.
4. Move all the variables to be included into the **Item** box.
5. Make sure that you select **Split-half** under the **Model** menu
6. Click **Statistics**.
7. Make sure that you select **Hotelling's T-square** and **Tukey's test of additivity** (see Figure 17.3.)
8. Click **Continue**.
9. Click **OK**.

Once you have clicked **OK**, SPSS will produce four boxes of information. The first box is just the case descriptive statistics (see Figure 17.4).

From this box we can see that 281, or 93.7% of the participants, completed all of the items included in the analysis, while 19 did not complete at least one item.

The next box contains the most important information, the reliability coefficients themselves (see Figure 17.5).

From this box we can see all of the information we really need. However, we are most interested in the Spearman-Brown coefficient (unequal length, as we have seven items in one group and six items in the other) and the Guttman split-half coefficient. We should remember from the earlier chi-square test that, as we do not have equal variance, we are going to use the Guttman split-half coefficient. In this example, we have a strong reliability coefficent (.79). Had we found equal variance, we could have used the Spearman-Brown coefficient.

FIGURE 17.3

Case Processing Summary

		N	%
Cases	Valid	281	93.7
	Excluded[a]	19	6.3
	Total	300	100.0

[a] Listwise deletion based on all variables in the procedure

FIGURE 17.4

Reliability Statistics

Cronbach's Alpha	Part 1	Value	.789
		N of Items	7[a]
	Part 2	Value	.632
		N of Items	6[b]
	Total N of Items		13
Correlation Between Forms			.610
Spearman-Brown Coefficient	Equal Length		.758
	Unequal Length		.759
Gunman Split-Halt Coefficient			.756

[a] The Items are: self educator, self musician, self perform, self ME, self artist, self musictch. self music scholar.
[b] The Items are: self ensemdir, self compose. self conductor, self ntertainer, self entrep, self teacher.

FIGURE 17.5

Notice also at the bottom of the box, SPSS lets you know which variables were included in each group.

The next box includes Tukey's test for nonadditivity, which tests the assumption of additivity (see Figure 17.6).

From this box, we can see that we violate the assumption of additivity, meaning that the variables included in the analysis have multiplicative interactions. Put simply, there is no simple linear relationship between the items on the instrument. This state is not uncommon in complex research studies. It should be reported, but should not necessarily stop you from continuing your analysis unless your analysis is based in simple observable behavior.

The final box provided by SPSS is the Hotelling's T^2. Remember that Hotelling's T^2 tells the researcher if the two groups have equal means, *not variance*. Keep in mind that this statistic is rarely reported and can be difficult to meet (especially given two different-sized groups, as in this example). If the reliability coefficient is high, the reliability analysis is robust against a violation of this assumption. See Figure 17.7.

From Figure 17.7, we can see that a significant difference ($p < .001$) exists between the two groups and we have violated the assumption of equal forms. This result is not surprising in this case, however, as we have unequal halves. Given the relatively high reliability coefficient (.789 as found in Figure 17.5) despite this finding, the researcher can feel safe continuing with an analysis.

ANOVA with Tukey's Test for Nonadditivity

			Sum of Squares	df	Mean Square	F	Sig
Between People			1613.674	275	5.868		
Within People	Between Items		3465.334	12	288.820	369.485	.000
	Residual	Nonadditivity	13.302[a]	1	13.302	17.100	.000
		Balance	2566.248	3299	.778		
		Total	2579.550	3300	.782		
	Total		6045.385	3312	1.825		
Total			7659.059	3587	2.135		

Grand Mean = 4.72
[a] Tukey's estimate of power to which observations must be raised to achieve additivity = 1.436.

FIGURE 17.6

Hotelling's T-Squared Test

Hotelling's T-Squared	F	df1	df2	sig
1660.243	132.918	12	269	.000

FIGURE 17.7

EXAMPLE 17.2 POSSIBLE SPLIT-HALF RELIABILITY REPORT

In order to assess the reliability of a scale designed to determine the occupational identity of preservice music educators ($N = 281$), the researcher conducted a split-half reliability analysis. Prior to the analysis, the researcher assessed the goodness of fit (equal variance) by using the chi-square test of parallel models and found unequal variance ($\chi^2 = 1546.33$, $p = <.001$). Because of this violation, the researcher employed the Guttman reliability coefficient, which is robust against this violation. The researcher also assessed the assumption of additivity and equality of means by using Tukey's test for nonadditivity and Hotelling's T^2. Despite violations of both of these assumptions (Tukey $F = 84.16$, $p = <.001$; Hotelling's T^2, $F = 132.92$, $p = <.001$), the researcher found a strong internal consistency of the occupational identity research instrument (.78).

Reporting a Split-Half Reliability Analysis

Given the dataset in our running example, one possible research report of the data may look like Example 17.2.

Split-Half Reliability-Analysis Writing Hints

From the research report in Example 17.2, we see some hints that might help you state your findings as clearly as possible while providing all of the necessary information.

- Make sure that you provide the following in your report:
  - N
  - Tukey's test of additivity, including F and p
  - Hotelling's T^2, including F and p
  - The chi-square test of parallel models, including the chi-square statistic and p
  - Discussion of the appropriateness of subsequently using the instrument
  - The strength of the internal consistency (is it high, moderate, or weak?)
- Per American Psychological Association guidelines, use first-person pronouns for clarity, especially when describing actions you took, while keeping all of your actions in past tense.
- Use the active voice.
- Avoid personification.
- Make sure all statistical indicators are in italics.
- Keep your actions as researcher in the past tense.

Split-Half Reliability Exercises

Given the second provided dataset, Split Half Example 2 ⏵, compute all of the requisite assumption tests, as well as the reliability analysis. The columns included in this dataset are a scale of 13 occupational identity issues in which participants indicated the extent to which they believe others saw them in different occupational roles. The columns include the following:

- Others see me as an educator.
- Others see me as a musician.
- Others see me as a performer.
- Others see me as a music educator.
- Others see me as an artist.
- Others see me as a music teacher.
- Others see me as a music scholar.
- Others see me as an ensemble director.
- Others see me as a composer.
- Others see me as a conductor.
- Others see me as an entertainer.
- Others see me as an entrepreneur.
- Others see me as a teacher.

Once you have computed the most appropriate statistical tests, write a report that includes all of the information required for a reader to assess the accuracy of your work as well as the possible implications that may exist.

IV | Nonparametric Tests

18 | Chi-Square

Use of Chi-Square Tests

As previously discussed in this book, a music education researcher may be required to use a nonparametric test should her data not meet one or more of the assumptions of the equivalent parametric tests. Additionally, however, a researcher may wish to explore the differences, based on categorical data, between two groups. In such a case, no parametric test would be appropriate. The chi-square (χ^2) test, however, would be.

As seen in Table 1.6 in Chapter 1, researchers using a chi-square test are hoping to see if any differences exist between two groups or two data points based on a categorical variable.

For example, a music education researcher may be interested in finding out if differences exist between string instrumentalists and wind instrumentalists in regard to their personality type as evaluated on a binary personality scale (e.g., Type A/Type B personality). Because both of the variables are categorical—a person is either a string player or a wind player and a person has either a Type A or Type B personality—no parametric test would be appropriate. The researcher would need to employ a chi-square test.

When music education researchers use a chi-square test, both variables must be categorical. The categories can, however, have more than two levels. That is, the categories do not need to be binary as in the example above. A researcher could use a chi-square test to explore differences between four categories of musicians (i.e., string players, wind players, percussionists, and vocalists) and three categories of hair color (i.e., blond, brunette, redhead).

In Example 18.1, we can see that music education researchers used chi-square analyses to determine if differences existed between categorical variables. In this example, the researchers categorized participants' responses into three categories of impact on music programs: highly positive/positive, no effect, negative/highly negative. They also compared these categorical variables to the categorical variables of school location (rural, urban, suburban) and school socioeconomic status (based on the percentage of free and reduced lunches).

> **EXAMPLE 18.1 EXAMPLE OF THE USE OF CHI-SQUARE ANALYSIS.**
>
> Chi-square analyses were used to test for differences in the ways principals perceived 10 variables to be impacting music programs: highly positive/positive, no effect, and negative/highly negative. Results indicated significant differences for each of the 10 variables ($p < .05$). Of the variables, four (budgeting, scheduling, No Child Left Behind, and standardized tests) were perceived to be negatively impacting music programs in more than 25% of schools. These variables were further analyzed to test for differences in responses by school location (rural, urban, suburban) and by school socioeconomic status (free and reduced-price lunch percentages). Results of chi-square analyses revealed that there were no significant differences ($p > .05$) for any of the four variables by location or socioeconomic status. In other words, school SES and location were not associated with principal perceptions of variables negatively impacting a given music program.
>
> ---
>
> Abril, C. R., & Gault, B. M. (2008). The state of music in secondary schools: The principal's perspective. *Journal of Research in Music Education, 56*(1), 68–81.

Assumptions of Chi-Square Tests

The chi-square test is relatively free of assumptions. Given the nonparametric nature of the test, the only real assumptions of the test are the following:

1. The data used should be either ordinal or nominal. Remember that nominal data is that which is purely categorical (i.e., which instrument a participant plays, the type of music a participant prefers most, etc.). Ordinal data is data that is placed in order, but can be interpreted as both categorical (as we do not know the real difference between agree and strongly agree, as we would with a scale) and continuous (as it could be interpreted as on a continuum of sorts). For the purposes of the chi-square test, the ordinal data is treated as categorical data.
2. You need two variables with at least two levels each.
3. You should have a relatively large sample size; otherwise, the chi-square test results can be misleading.

Meeting the Assumptions of the Chi-Square Test

Meeting the assumptions of the chi-square test is relatively easy and is most commonly completed by the design of the study and the types of variables collected.

Research Design and the Chi-Square Test

There is really no experimental research design associated with the chi-square test. Given that the test is designed to examine two or more categorical variables, it is most commonly used in survey designs in which participants provide demographic data and responses to other categorical or ordinal data questions.

The Null Hypothesis for the Chi-Square Test

As with many other statistical procedures, the null hypothesis for the chi-square test is that the means for all the groups and comparisons will be equal. It could generically be stated as

$H_0 = \mu_1$ is *independent* of or not associated with μ_2

For our musical example above of comparing string and wind instrumentalists and personality types, the null hypothesis could be stated as

$H_0 = \mu_{\text{instrument type}}$ is independent of or not associated with $\mu_{\text{personality type}}$

The alternative hypothesis would be that the two variables are not independent of each other and could be stated as

$H_a = \mu_1$ is *not independent* of or is associated with μ_2

Therefore, in our musical example, the alternative hypothesis could be stated as

$H_a = \mu_{\text{instrument type}}$ is not independent of or is associated with $\mu_{\text{personality type}}$

Setting Up a Database for the Chi-Square Test

Setting up a database for a chi-square test is quite simple. It will require two columns, each containing a categorical variable that has been coding into numbers. So, in our musical example, the string players have been coded as 1s and the wind players have been coded as 2s. Additionally, Type A people have been coded as 1s, while Type B personalities have been coded as 2s (see Figure 18.1).

Conducting a Chi-Square Test in SPSS

For this section, please use the provided example dataset (Chi-square Example 1 ▶) so that you can complete the same analysis you will be reading in order to help you better understand how to conduct and analyze the data through doing.

	InstrumentType	PersonalityType
1	1	1
2	2	1
3	1	1
4	2	1
5	1	1
6	1	1
7	1	1
8	1	1
9	2	2
10	2	2
11	2	2
12	2	2
13	2	1
14	2	1
15	1	2

FIGURE 18.1

In order to complete the test, follow these steps:

1. *Click* **Analyze**.
2. *Click* **Descriptive Statistics**.
3. *Click* **Crosstab**s.
4. *The window that appears will ask you to place one test item in the* **Row(s)** *box and one test item in the* **Column(s)** *box (see Figure 18.2).*
5. *Click* **Statistics**.
6. *Make sure to select the* **Chi-square** *box as well as* **Phi and Cramer's V** *if your data is nominal or* **Somer's d** *if your data is ordinal. In this case, our data is nominal, so we select* **Phi and Cramer's V** *(see Figure 18.3).*
7. *Click* **Continue**.
8. *After you click* **Continue,** *on the next screen, click* **Cells**.
9. *Make sure to select* **Observed** *in the* **Counts** *box and that you click* **Row**, **Column**, *and* **Total** *in the* **Percentages** *box (See Figure 18.4).*
10. *Click* **Continue**.
11. *Before clicking* **OK** *to run the test, make sure to select the* **Display clustered bar charts** *box. This can be a very useful tool in interpreting the findings.*
12. *Click* **OK**.

Once you click **OK**, you will receive five different boxes of information. The first box is just the process summary table (see Figure 18.5). In this box, we are informed as to the *N* included in the analysis as well as any missing cases. In this example, all 137 participants had both required data points and so were included in the study.

The second box provided by SPSS is the cross-tabulation box (see Figure 18.6). This table provides the descriptive statistics for the cross-tabulation. Each percentage that we asked for (row, column, and total) are interpreted differently. For example, in Figure 18.6 (keeping in mind that 1 = strings and 2 = winds, and that 1 = Type A and 2 = Type B), we can see that

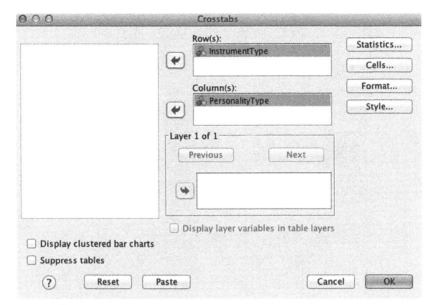

FIGURE 18.2

FIGURE 18.3

73.8% of string players are Type A personalities. This is our row percentage. We can further see that 51.1% of the participants in this study who are Type A personalities are string players. Finally, we can see that overall in the study, 32.8% of the participants in the study are string players and Type A personalities. To take a look at another of these to double-check for understanding, if

FIGURE 18.4

Case Processing Summary

	Cases					
	Valid		Missing		Total	
	N	Percent	N	Percent	N	Percent
Instrument Type* Personality Type	137	100.0%	0	0.0%	137	100.0%

FIGURE 18.5

Instrument Type* Personality Type Cross tabulation

			Personality Type		
			1	2	Total
Instrument Type	1	Count	4	16	61
		% within Instrument Type	73.8%	26.2%	100.0%
		% within Personality Type	51.1%	32.7%	44.5%
		% of Total	32.8%	11.7%	44.5%
	2	Count	43	33	76
		% within Instrument Type	56.6%	43.4%	100.0%
		% within Personality Type	48.9%	67.3%	55.5%
		% of Total	31.4%	24.1%	55.5%
Total		Count	88	49	137
		% within Instrument Type	64.2%	35.8%	100.0%
		% within Personality Type	100.0%	100.0%	100.0%
		% of Total	64.2%	35%	100.0%

FIGURE 18.6

you try to interpret the data regarding wind players who are Type B personalities, you would see that

- 43.4% of wind players are Type B personalities (row percentage).
- 67.3% of Type B personalities are wind players (column percentage).
- 24.1% of all participants are both Type B personalities and wind players (total percentage).

The third box created by SPSS is the actual chi-square test outcome (See Figure 18.7). Although several points of information are provided in this box, we are really interested only in the Pearson chi-square line. In Figure 18.7, we see that the chi-square statistic for this test is 4.35 with 1 degree of freedom and a significance level of .037. In this case, the difference between the two groups is significant.

The next box provided is the symmetric measures table (Figure 18.8), which provides the phi (Φ) and Cramer's V. These statistics serve as effect-size analysis for a chi-square test. Generally speaking, a phi and Cramer's V between 0.1 and .29 is small. A phi and Cramer's V between .30 and .49 is medium, and a phi and Cramer's V at .5 or above is a large effect size. It is most appropriate to report phi when the chi-square test is using a 2 × 2 table (i.e., there are two

Chi-Square Tests

	Value	df	Asymptotic Significance (2-sided)	Exact Sig. (2-sided)	Exact Sig. (1-sided)
Pearson Chi Square	4.353[a]	1	.037		
Continuity Correction[b]	3.637	1	.057		
Likelihood Ratio	4.422	1	.035		
Fisher's Enact Test				.048	.028
Linear-by-Linear Association	4.321	1	.038		
N of valid Cases	137				

[a] 0 cells (0.0%) have expected count less than 5. The minimum expected count is 21.82.
[b] Computed only for a 2 x 2 table

FIGURE 18.7

Symmetric Measures

		Value	Approximate Significance
Nominal by Nominal	Phi	.178	.037
	Cramer's V	.178	.037
N of Valid Cases		137	

FIGURE 18.8

groups with two levels, as in our example. Computing phi is a relatively easy thing to do without the assistance of any software. The formula for phi is

$$\varphi = \sqrt{\frac{x^2}{n}}$$

In this formula, n is the number of observations. So in our example, we could enter the chi-square statistic of 4.35 and the n of 137. The equation would look like

$$\varphi = \sqrt{\frac{4.35}{137}} = .178$$

If, however, you employs a chi-square test with more than a 2 × 2 table, it is most appropriate to employ Cramer's V to interpret effect size. Also, the interpretation of effect size changes, given the size of the table being examined. Cramer's V builds upon the formula for phi, but takes into consideration different degrees of freedom on the basis of the size of the table. In order to know how many degrees of freedom a chi-square table has, simply take the number of rows minus 1 and multiply it by the number of columns minus 1. So in our 2 × 2 table of our example, we would have only one degree of freedom because

$$2 - 1 \times 2 - 1 = 1$$

This is why the statistics for phi and Cramer's V are the same in our 2 × 2 example. If, however, we had a 2 × 3 table, we would find that phi and Cramer's V would be different. For example, let's say that a music researcher was interested in any association between two personality types and string players, wind players, and percussionists. The rest of the chi-square process would be the same, but we would need to use Cramer's V to understand effect size. The formula for Cramer's V is

$$V = \sqrt{\frac{x^2}{n \times df}}$$

As we can see, the formula is practically the same as phi, but takes into account different degrees of freedoms. Because of this latitude, we need to be more careful as to how we interpret the Cramer's V outcome. With only 1 degree of freedom, V is interpreted the same as phi (i.e., 0.1 and .29 is small, .30 and .49 is medium, .5 or above is a large effect size). However, as the degrees of freedom increase, you need to divide these guidelines by the square root of the degrees of freedom. For example, if you had 2 degrees of freedom, we would alter the .5 requirement for a large effect size thus:

$$\text{Square root of } 2(df) = 1.41$$
$$\text{Large effect size originally set at } .50$$
$$.50 \div 1.14 = .35$$

The new criterion for a large effect size would be .35 and can be easily calculated for any given degrees of freedom. For example, should the chi-square test have 7 degrees of freedom, the new standard for a large effect size could be calculated as

$$\text{Square root of } 7(df) = 2.65$$
$$\text{Large effect size originally set at } .50$$
$$.50 \div 2.65 = .19$$

It may seem odd that the requirement for a large effect size is getting lower rather than higher. However, keep in mind that the more associations being explored in a chi-square test, the more challenging they are to find.

Unfortunately, the effect sizes of chi-square tests are rarely reported in music education research, but should be, especially if the chi-square test is the primary means used to interpret the data.

The final box that we asked SPSS to create is the bar chart (see Figure 18.9). This chart gives us a visual means to interpret the data and can help clarify the data for our readers and ourselves. As seen in Figure 18.9, relatively equal numbers of string players and wind players are Type A personalities. However,

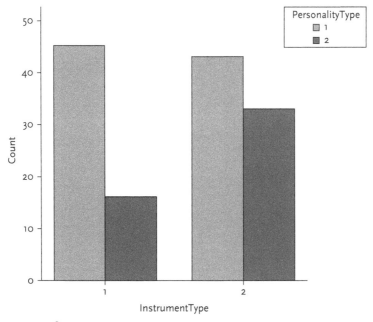

FIGURE 18.9

EXAMPLE 18.2 POSSIBLE CHI-SQUARE TEST REPORT OF THE EXAMPLE DATA SET

In order to explore the possible association between family of instrument played and personality type, I had participants (N = 137) respond to a questionnaire regarding what instrument they played and complete a personality test, which resulted in the participants being categorized as either a Type A personality or a Type B personality. Sixty-one of the participants played a string instrument, while the remaining played a wind instrument (n = 76). The majority of participants (n = 88) were identified as Type A personalities. Nearly three-quarters of the string players (73.8%) had a type A personality, while only half (56.6%) of the wind players had a Type A personality. I found a significant association between instrument played and personality type (χ^2 = 4.35, df = 1, p = .037). String players were much more likely to be a Type A personality. Despite the statistically significant association, little practical significance can be assumed on the basis of a relatively low effect size (Φ = .18).

on the basis of just one visual interpretation, it seems pretty clear that more wind players in this study were Type B personalities than string players.

Reporting a Chi-Square Test

Once you have completed all of these steps, you have all of the information you need to complete a research report of a chi-square analysis in a thesis, dissertation, or journal setting.

One possible report of the data found in our running example in this chapter may look like the one found in Example 18.2.

Chi-Square Writing Hints

From the report in Example 18.2, we can glean some general writing hints for when reporting a chi-square test.

- Make sure you provide in your report:
    - N
    - Any n's
    - χ^2
    - df
    - p
    - Φ (phi) or Cramer's V
    - Any descriptive statistics that will help the reader better understand the analysis.

- Per American Psychological Association guidelines, use first-person pronouns for clarity, especially when describing actions you took, while keeping all of your actions in past tense.
- Use the active voice.
- Use first-person pronouns
- Make sure all statistical indicators are italicized.
- Keep your actions as researcher in the past tense.

Chi-Square Exercises

Given the second provided data set (Chi-square Example 2 ▶), compute the necessary analyses in SPSS.

In this example, the two columns are the following:

- Highest earned music education degree of in-service music educators (1 = bachelor's degree, 2 = graduate degree)
- Level of comprehensive music education (i.e., composition, improvisation, historical context, etc.) employed in their classrooms in two categories (1 = low level, 2 = high level).

In this study, a music education researcher was curious to see if an association existed between the educational level of music teachers and how much they engaged in nontraditional performance activities in their classroom. Do those with a graduate education feel more comfortable using these methods as one might think?

Once you have computed the most appropriate statistical test, write a report that includes all of the necessary information for readers to understand and trust the reported statistics.

19 | Mann-Whitney U Test

Use of the Mann-Whitney U Test

Much like an independent-samples t-test, music education researchers can use the Mann-Whitney U test when they want to compare the means of two groups and their data does not quite conform to typical parametric standards. Researchers also use this nonparametric version of the independent-samples t-test when their data is ordinal rather than scaled (although the results are rarely different in these cases). As with the independent samples t-test, the independent variable is usually a grouping or level of a variable, while the dependent variable is a continuous variable that has been derived from some researcher instrument or observation.

As seen in Table 1.6 in Chapter 1, music education researchers using a Mann-Whitney U test are seeking to answer a simple research question: does a differences exist between two groups' means on one dependent variable?

For example, a music education researcher may be interested in comparing the university music-faculty course evaluations completed by male and female students to see if a gender bias in responses may exist. Do students who self-identify as male rate their music professors differently from the way those who self-identify as female do? However, upon close inspection of the descriptive data, the researcher may have discovered that the collected data does not seem to be normally distributed (a requirement for a parametric test). Or the researcher may make the logic-based argument that the responses on the faculty evaluation sheets are ordinal and the nonparametric version of the statistical procedure may be more appropriate and parsimonious. In this example, the evaluation scores serve as the dependent variable, while the binary grouping based on self-identified gender serves as the independent variable. Thus, whenever a music education researcher employs a Mann-Whitney U test, the dependent variable must be either ordinal or continuous, while the independent variable must be categorical with not more than two levels or groups.

In Example 19.1, we see that a music education researcher has employed a Mann-Whitney test to find if significant differences existed between two

> **EXAMPLE 19.1 MANN-WHITNEY U USED TO DETERMINE DIFFERENCES BETWEEN TWO GROUPS**
>
> Mann-Whitney U tests performed on the frequency of contour-preserving and contour-violating errors with age as the grouping variable revealed significant differences between the age groups in terms of the frequency of contour-violating errors ($U = 88.5$; $Z = -2.21$; $p = .03$). The younger group made significantly more contour-violating errors than did the older group. However, the frequency of contour-preserving errors did not differ significantly between age groups.
>
> ---
>
> Gudmundsdottir, H. R. (2010). Pitch error analysis of young piano students' music reading performances. *International Journal of Music Education, 28*(1), 61–70.

different age groups on their ability to maintain melodic contour when identifying musical errors. In this example, the researcher found that age group did impact how often participants violated the melodic contour, but found no difference between the two group's abilities to identify errors when the melodic contour was correct.

Assumptions of the Mann-Whitney U Test

The Mann-Whitney U test has three primary assumptions, which mirror, to some extent, those of the independent-samples t-test:

1. The independent variable (or grouping variable) is bivariate. That is, the independent variable is membership in one or another group or category.
2. The data collected for the dependent variable is either ordinal or continuous rather than categorical.
3. Each observation is independent of any other observation (most often accomplished through random sampling).

Moreover, as with the t-test, it is easier to mitigate violations of these assumptions if you have equal numbers of participants or data points in each group and if there are more than 30 data points in each group. It is also helpful if both groups have relatively equal distributions, although having unequal distributions is not fatal to the test.

Meeting the Assumptions of a Mann-Whitney U Test

Meeting the assumptions of the Mann-Whitney U test is relatively easy and can be successfully done primarily through the design of your research study, as the assumptions are mostly about what kinds of data are analyzed. If the researcher collected the correct types of data, the primary assumptions of the test will be met.

Research Design and the Mann-Whitney U Test

The Mann-Whitney U test can be used to compare any two groups that are independent of each other. So, the possible research designs that utilize this type of inferential test primarily include

(A) X   O
    ―――
         O

(B) X   O   X
    ―――――
         O   O

Let's take a look at a few examples of how a music education research may use the Mann-Whitney test in these designs.

(A) $X_{\text{professional development seminar on technology in the classroom}}$ $O_{\text{self-reported comfort using technology}}$
―――――――――――――――――――――――――――――――――
                                                    $O_{\text{self-reported comfort using technology}}$

In the example above, two groups of teachers were examined, one group participated in a professional development seminar regarding the use of technology in the classroom. The other group had no professional development. The researcher was interested in whether or not the two different groups of teachers reported a different level of comfort with using the technology in the classroom. As stated above, this research design creates a perfect opportunity to use the Mann-Whitney U test. The dependent variable is an ordinal variable (i.e., self-reported level of comfort using technology on a 5-point scale) and the independent variable is bivariate (i.e., music teachers who received professional development and music teachers who did not receive professional development).

(B) $O_{\text{use of non-western musics}}$ $X_{\text{course in non-western musics}}$ $O_{\text{use of non-western musics}}$
―――――――――――――――――――――――――――――――――
     $O_{\text{use of non-western musics}}$ $O_{\text{use of non-western musics}}$

In this example, a music education researcher is interested in whether or not taking a graduate-level course in non-western music increases music teachers' use of non-western musics in their classrooms. Also in this example, the researcher has several options for analysis. If the data did not meet the assumption of normality for the t-test, the researcher could use the Mann-Whitney U to evaluate if differences existed between the two groups of teachers at the time of the pre-test or at the time of the post-test. More simply, however, the researcher could also use the Mann-Whitney U test to see if any differences existed between the two groups' mean gain scores.

Note, however, that the Mann-Whitney U test does not necessarily need to have a specific experimental research design. Researchers who use survey designs will often employ this test to examine differences between two groups of respondents to their surveys. Given the extensive use of ordinal data (i.e., Likert-type scales) in questionnaires, survey researchers can heavily rely on Mann-Whitney U tests to help make greater sense out of the data they collect.

The Null Hypothesis for Mann-Whitney U Test

The null hypothesis for the Mann-Whitney U test is similar to that of the independent-samples t-test:

$H_0 = \mu_1 = \mu_2$

For our musical example above regarding comfort with technology, the null hypothesis would be that no difference exists in comfort using technology between the teachers who participated in professional development and those who did not. It could be stated as

$H_0 = \mu_{\text{music teachers receiving professional development}} = \mu_{\text{music teachers not receiving professional development}}$

The alternative hypothesis would be that a difference did exist between the two groups of teachers and could be stated as

$H_a = \mu_1 \neq \mu_2$

Or, more specific to our musical example,

$H_a = \mu_{\text{music teachers receiving professional development}} \neq \mu_{\text{music teachers not receiving professional development}}$

Finally, the researcher may be interested in making a directional hypothesis, which would be stated as

$H_1 = \mu_1 < \mu_2$

If our music researcher has reason to believe, on the basis of previous research or other evidence, that a directional hypothesis was appropriate and that it is likely that those who received professional development would be more comfortable, the directional hypothesis could be stated as

$H_1 = \mu_{\text{music teachers receiving professional developement}} > \mu_{\text{music teachers not receving professional development}}$

Setting Up a Database for a Mann-Whitney U Test

The database for a Mann-Whitney U test will have two required columns. In one column we can find the grouping variable. This column should have only two possible inputs, as only two groups or levels of a variable can be used as the

	SIGENDER	FACEVAL
1	1	2
2	1	3
3	1	4
4	1	3
5	1	2
6	1	1
7	1	2
8	1	3
9	1	3
10	1	2
11	1	3
12	1	4
13	1	3
14	1	2
15	1	1

FIGURE 19.1

independent variable in this test. The other column should have the dependent variable, which should be ordinal or continuous. In our example (see Figure 19.1) the independent variable is that of self-identified gender, which has been coded as males as 1 and females as 2. The dependent variable is the participants' rating (on a 5-point scale) on a music faculty members' course evaluations. The researcher is interested to see if a difference based on the gender of the students exists.

Conducting the Mann-Whitney U Test in SPSS

For this section, please use the provided example database (MannWhitney 1 ▶ in order to follow along through the process while you are reading.

In order to conduct the Mann-Whitney U test in SPSS, follow these steps:

1. Click **Analyze**.
2. Click **Nonparametric Tests**.
3. Click **Legacy Dialogs**.
4. Click **Two-Independent-Samples Test**.
5. Move the dependent variable into the **Test Variable List**.
6. Move the independent variable into the **Grouping Variable** space.
7. Click the **Define Groups** button.
8. Indicate that group 1 will be represented by the value of 1 and group 2 will be represented by the value of 2. All you are doing here is telling SPSS how you labeled your two groups (see Figure 19.2).
9. Make sure to select **Mann-Whitney U** under **Test Type**.
10. Click **Options** and make sure to select the **Descriptives** box.
11. Click **OK**.

Once you have clicked **OK**, you will get an output with three boxes. The first box is important, as it provides the descriptive statistics for reporting and

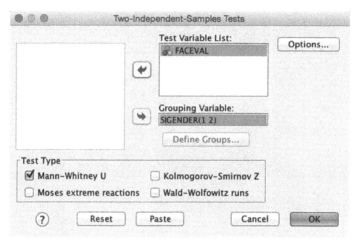

FIGURE 19.2

Descriptive Statistics

	N	Mean	Std. Deviation	Minimum	Maximum
FACEVAL	82	3.45	1.124	1	5
SIGENDER	82	1.51	.503	1	2

FIGURE 19.3

analysis (see Figure 19.3). From this box, we can see the N, mean, standard deviation, and range of the two groups being examined. On the basis of this information, we can see that the mean faculty evaluation score was 3.45 and that slightly fewer men were included in the study.

The second box (see Figure 19.4) provides the mean ranks found in the data. We are able to use the mean-ranks data to interpret which group scored faculty higher or lower. In this instance the mean rank for the men's scores was 27.64, while the mean rank for the women's scores was 54.70. From this information we can interpret that women scored their faculty members higher than men scored faculty members. What we do not yet know is if that difference is statistically significant.

The final box created by SPSS gives us the actual statistical outcome of the Mann-Whitney U test (see Figure 19.5). In this box, we see that the actual U statistic is 285.5 and that this is a significant statistical difference (.000 or $p < .001$). This box also reports Z, which will be important later for calculating effect size.

Is this enough information? Not quite yet. Although many researchers stop at reporting the statistical significance of Mann-Whitney U test, more information would be most helpful to determine the practical significance of the finding. For the Mann-Whitney test, SPSS does not create a report of effect size.

Ranks

	SIGENDER	N	Mean Rank	Sum of Ranks
FACEVAL	1	40	27.64	1105.50
	2	42	54.70	2297.50
	Total	82		

FIGURE 19.4

Test Statistics[a]

	FACEVAL
Mann-Whitney U	285.500
Wilcoxon W	1105.500
Z	−5.335
Asymp. Sig. (2-tailed)	.000

[a] Grouping Variable: SIGENDER

FIGURE 19.5

However, computing the effect size (r) for this test is quite simple. This is a straightforward formula that will take you only a few moments to compute, but will give your analysis much greater depth and meaning. The formula for r (effect size for Mann-Whitney U) is

$$r = \frac{Z}{\sqrt{N}}$$

In this formula, we obtain Z from the third box provided in the SPSS output (See Figure 19.5) and we know that the N of the analysis is 82, based on the first box provided by SPSS (see Figure 19.3). With this information, we can begin to calculate r.

$$r = \frac{-5.33}{\sqrt{82}}$$

$$r = \frac{-5.33}{9.06}$$

$$r = -.59$$

Do not be concerned about whether or not the outcome is positive or negative; we are concerned about the magnitude of the difference between the two groups at this point.

In general, Mann-Whitney U tests with an r of .1 to 2.9 are small effect sizes, .3 to .49 are moderate effect sizes and .5 and above are large effect sizes. So, in our example, we have a rather large effect size and can therefore argue for practical significance of our findings.

> **EXAMPLE 19.2 POSSIBLE MANN-WHITNEY U TEST RESEARCH REPORT OF THE EXAMPLE DATA**
>
> In order to determine if a difference existed between students who self-identified as males and those who self-identified as females in regard to their evaluations of their music faculty members, we employed a Mann-Whitney U test. We employed this nonparametric test because of the ordinal nature of the data used for the dependent variable. We found that females ($n = 42$, mean rank = 54.70) scored their music professors higher than their male colleagues ($n = 40$, mean rank = 27.64) ($U = 285.5$, $Z = -5.34$, $p < .001$, $r = -.58$). On the basis of the effect size of this difference, we argue that this finding has practical significance as well as statistical significance.

Reporting a Mann-Whitney U Test

Once you have conducted the Mann-Whitney U test and computed the effect size, you have enough information to write a full report of your findings. Given the information we found in our running example, one possible report of the data may look like the one found in Example 19.2.

Mann-Whitney U Test Writing Hints

From Example 19.2 we see a few writing hints that may help you state your findings as directly as possible.

- Make sure that you report the following:
  - Descriptives
  - N of each group
  - The mean ranks
  - U
  - Z
  - p
  - r
  - Discussion of practical and statistical significance
- Per American Psychological Association guidelines, use first-person pronouns for clarity, especially when describing actions you took, while keeping all of your actions in past tense.
- State your findings in the active voice (i.e., I or we did X or the participants did X).
- State your findings in the past tense.
- When you are reporting statistical information, make sure that the label of the statistic (i.e., U, Z, p, r, etc.) is in italics.

- Make sure to report Z, as it helps the reader trust your effect size calculation.
- Always overtly state the type of statistical procedure you are using (e.g., Mann-Whitney U test). Even if the well-informed reader can infer the type of test, it is helpful to overtly report it.
- Many readers will appreciate a quick discussion as to why you employed the nonparametric version of the test.

Mann Whitney U Test Exercises

Given the second dataset provided, MannWhitney Example 2 ⓘ, complete the required analyses. In this example, the two columns are the following:

- Grouping variable—group 1 includes music teachers who completed a week-long professional development seminar on the use of technology in the music classroom. Group 2 includes music teachers who did not have the opportunity to complete the seminar.
- Dependent variable—participants' response to a question regarding how comfortable they are using technology in their music classrooms on a scale of 1 to 5 (1 = not comfortable at all, 5 = very comfortable).

In this study, the researcher was interested in finding out if the week-long seminar was useful in alleviating or mitigating discomfort with employing technology in the music classroom, not skill or frequency of technology use.

Once you have completed the analysis, write a research report that includes all of the information necessary for a reader to understand and trust your analysis.

20 | Kruskal Wallis H Test

Use of the Kruskal Wallis H Test

The Kruskal Wallis H test is the nonparametric counterpart to the ANOVA. Therefore, researchers can use the Kruskal Wallis H test when they need to compare the means of two or more groups or levels of a variable and their data does not meet the assumptions of an ANOVA. Keep in the mind that if you have only two groups or levels of a variable (e.g., comparing high-string players to low-string players in your orchestra) the more appropriate nonparametric test would be the Mann-Whitney U test. The most simple and parsimonious version of a statistical procedure, even the nonparametric ones, is still almost always the best.

As seen in Table 1.6 in Chapter 1, researchers using the Kruskal Wallis H test are seeking to answer a simple question: does a difference exist between two or more groups on one dependent variable?

For example, a music education researcher may be interested in exploring differences in string students' vibrato skills among those who are rising into high school from three different feeder schools. Because there are more than two groups, the Mann-Whitney U test is no longer an option, so the researcher would need to employ a Kruskal Wallis H test. As with the independent sample t-test and ANOVA, you could think of the Kruskal Wallis H test as an extension of the Mann-Whitney U test. Whereas the Mann-Whitney U test can be used with only two groups or levels of a variable, a Kruskal Wallis H test can explore the differences among three or more groups or levels of a variable.

When music education researchers use the Kruskal Wallis H test to examine the differences among three or more groups, the independent variable (*must be categorical*) is the different groups in the model and the dependent variable is a score (*must be ordinal or continuous*) on some research instrument.

In Example 20.1 we can see that researchers used the Kruskal Wallis H test to examine the differences between three groups with different levels of experience with authentic-context-learning activities and their initial teaching skill.

> **EXAMPLE 20.1 KRUSKAL WALLIS H TEST USED TO DETERMINE DIFFERENCES AMONG THREE GROUPS**
>
> Kruskal Wallis analysis of variance (ANOVA) was used to determine whether there was a significant difference in the initial teaching performance among those with high, medium, and low levels of involvement in authentic context learning (ACL) activities in instrumental music settings. An overall level of involvement in ACL activities was determined by totaling the scores of the four ACL activities for each subject. Using naturally occurring gaps in a linear display of the overall ACL activity scores, we categorized the subjects into three groups, one representative of a low level of ACL involvement ($n = 16$), another representative of a medium level of ACL involvement ($n = 9$), and a third group representative of a high level of ACL involvement ($n = 5$). Significant differences in initial teaching ability were found among the three groups ($H = 13.06$, $df = 2$, $p = .001$).
>
> ---
>
> Paul, S. J., Teachout, D. J., Sullivan, J. M., Kelly, S. N., Bauer, W. I., & Raiber, M. A. (2001). Authentic-context learning activities in instrumental music teacher education. *Journal of Research in Music Education, 49*(2), 136–145.

Although the authors in Example 20.1 did not overtly state why they elected to use the nonparametric version of the test, it seems clear that the reason was the small number of participants in each group. It would be unlikely that groups consisting of 16, nine, and five individuals would meet the assumption of a normally distributed data or homogeneity of variance, which are required for the parametric tests.

Post Hoc Tests for the Kruskal Wallis H Test

As with the ANOVA, the Kruskal Wallis H test does not give the reader enough information to interpret the findings. Because more than two groups are usually present in the analysis, post hoc tests are usually required to tell us where the differences actually exist. For the Kruskal Wallis H test, there are two ways post hoc tests can be completed. The first is to use Mann-Whitney U tests (see Chapter 19) as post hoc tests, as seen in Example 20.2

Another method for completing post hoc tests for the Kruskal Wallis H test is Dunn's multiple-comparison procedure. This may be the most parsimonious manner of analyzing the data from a Kruskal Wallis H test, as seen in Example 20.3.

Assumptions of the Kruskal Wallis H Test

The Kruskal Wallis H test has three primary assumptions that are generally met on the basis of the research design and types of data collected rather than

EXAMPLE 20.2 USE OF MANN-WHITNEY U AS POST HOC TEST FOR KRUSKAL WALLIS H TEST

We used the participants' teaching level as the independent variable and the response to the Likert-type scale in regard to frequency of modeling as the dependent variable. We found a significant difference in the frequency of modeling based on teaching level, $X^2(2) = 8.86$, $p = .01$. In order to find differences that existed between groups, we conducted follow-up Mann-Whitney U pairwise comparisons. Private teachers reported modeling significantly more often than K–12 teachers ($U = 8535$, $p = .005$), but not college teachers ($U = 7083$, $p = .035$).

Pellegrino, K., & Russell, J. A. (2015, Spring). String teachers' practices and attitudes regarding their primary string instrument in settings inside and outside the classroom. *Bulletin of the Council for Research in Music Education*, (204), 9–26.

EXAMPLE 20.3 USE OF DUNN'S MULTIPLE COMPARISONS AS POST HOC TESTS

Dunn's multiple comparison procedure was used to determine that a significant difference ($p < .05$) in teaching performance existed between subjects with high and low levels of ACL involvement and between subjects with high and medium levels of ACL involvement. No significant difference ($p < .05$) in teaching performance, however, was found between subjects with medium and low levels of ACL involvement

---

Paul, S. J., Teachout, D. J., Sullivan, J. M., Kelly, S. N., Bauer, W. I., & Raiber, M. A. (2001). Authentic-context learning activities in instrumental music teacher education. *Journal of Research in Music Education*, 49(2), 136–145.

any inherent condition of the data itself. The assumptions of the Kruskal Wallis H test are the following:

1. The data collected for the dependent variable should be either ordinal (e.g., some response to a Likert-type scale) or continuous.
2. The data collected for the independent variable is categorical (e.g., which instrument a student plays, which ensemble a student is in, etc.).
3. Each observation is independent of any other observation.

You may notice that the ANOVA had two additional assumptions, that of normal distribution and homogeneity of variance. The Kruskal Wallis H test is robust against those assumptions, thus the difference between the parametric ANOVA and the nonparametric Kruskal Wallis H test.

Meeting the Assumptions of the Kruskal Wallis H Test

As stated above, meeting the assumptions of the Kruskal Wallis H test is mostly a means of research design. If you collect data for a dependent variable that is ordinal or continuous, collect a categorical variable that is at least two groups or levels of a variable, and ensure that observations are independent of one another, you have met the assumptions of the test.

Research Design and the Kruskal Wallis Test

The Kruskal Wallis H test can be used in the same research designs as the one-way ANOVA. Therefore, as with the ANOVA, the possible research designs that utilize this type of inferential test include (O = an observation, X is a treatment of any given type):

(A) O X_1 O

 O X_2 O

 O X_3 O

 O X_4 O

(B) X_1 O

 X_2 O

         O

(C) O

 O

 O

Let's take a look at a few musical examples of what these designs might be in the area of music education.

(A) $O_{\text{vibrato pre-evaluation}}$ $X_{\text{control group}}$ $O_{\text{vibrato post-evaluation}}$

$O_{\text{vibrato pre-evaluation}}$ $X_{\text{1 minute of no instrument tasks}}$ $O_{\text{vibrato post-evaluation}}$

$O_{\text{vibrato pre-evaluation}}$ $X_{\text{5 minutes of no instrument tasks}}$ $O_{\text{vibrato post-evaluation}}$

$O_{\text{vibrato pre-evaluation}}$ $X_{\text{10 minutes of no instrument tasks}}$ $O_{\text{vibrato post-evaluation}}$

In the example design above, a music education researcher was interested in the impact of different lengths of time spent in working on vibrato exercises without instruments in students' hands. First, the researcher administered an initial vibrato test to see where students' skills existed prior to any treatment. The first group received no vibrato instruction (the control group). The second group received one minute of vibrato instruction per class. The third group of students received five minutes of vibrato instruction, while the fourth group received 10 minutes. However, because of a low enrollment, the researcher could not use the parametric ANOVA, as the data would not meet the assumptions. The researcher could use Kruskal Wallis H test to find out if there were any differences in mean gain scores. In this example, the students' mean gain score from first observation to second observation is the continuous dependent variable, while the categorical independent variable is group membership (i.e., members of treatment 1, 2, 3, 4). This design could be altered in many ways to include as few as two groups and, theoretically, as many groups as could be examined (although there are logical limitations to this test).

If the researcher uses an ANOVA in design B, she can examine the differences between any number of groups without a pre-test by using the control group as the comparison point.

(B) $X_{\text{high string students}}$ $O_{\text{error discrimination test}}$

$X_{\text{low string students}}$ $O_{\text{error discrimination test}}$

$O_{\text{error discrimination test}}$

In this example (B), a researcher was interested in whether or not playing a high string or low string instrument made any difference on one's ability to aurally identify musical errors in comparison to those with no string-playing experience. The researcher would be able to use the Kruskal Wallis H test to compare the error discrimination scores (dependent variable) of the participants in each group (independent variable), including students who did not participate in a formal string music ensemble in order to compare.

In design C, we see that no actual experimental design exists. Rather, researchers may use the Kruskal Wallis H test to examine the differences between groups that are based upon individual difference variables (those which were not manipulated or controlled by the researcher such as instrument family, level of education—bachelor's, master's, doctorate, etc.) This type of design is commonly found in descriptive studies as survey research.

(C) $O_{\text{public school teachers' career plans}}$

$O_{\text{private school teachers' career plans}}$

$O_{\text{private studio teachers' career plans}}$

In this example (C) a researcher could use a Kruskal Wallis H test to see if differences existed between three different groups on a survey of intended career plans (e.g., stay in their jobs, move to other similar jobs, or leave music teaching altogether). The researcher is looking to see if those with different music teaching positions have different future career plans.

The Null Hypothesis for the Kruskal Wallis H Test

The null hypothesis for the Kruskal Wallis H test is the same as that of the ANOVA: the means of all groups or levels of a variable will be equal. It could be stated as

$H_0: \mu_1 = \mu_2 = \mu_3 = \mu_4 = \mu_5$ etc.

For our musical example above regarding the impact of different lengths of vibrato instruction, the null hypothesis would be that the means of the different treatment groups would be statistically the same.

$H_0: \mu_{\text{control group}} = \mu_{\text{1 minute of no instrument tasks}} = \mu_{\text{5 minutes of no instrument tasks}}$
$= \mu_{\text{10 minutes of no instrument}}$

$H_0: \mu_{\text{control group}} = \mu_{\text{1 minute of no instrument tasks}} = \mu_{\text{5 minutes of no instrument tasks}}$
$= \mu_{\text{10 minutes of no instrument}}$

The alternative hypothesis would be that there would be a difference between the mean of each of the groups or levels. It would be stated as

$H_a: \mu_1 \neq \mu_2 \neq \mu_3 \neq \mu_4 \neq \mu_5$ etc.

In our musical example, the alternative hypothesis could be written as

$$H_a: \mu_{\text{control group}} \neq \mu_{\text{1 minute of no instrument tasks}} \neq \mu_{\text{5 minutes of no instrument tasks}} \neq \mu_{\text{10 minutes of no instrument}}$$

Finally, the directional hypothesis would state that the mean of one or more of the groups or levels of variable would not be equal and that one would be higher or lower than the other. This hypothesis would be stated as

$$H_1: \mu_1 < \mu_2 < \mu_3 < \mu_4 < \mu_5 \text{ etc.}$$

In our musical example, the directional hypothesis might be listed as

$$H_a: \mu_{\text{control group}} < \mu_{\text{1 minute of no instrument tasks}} < \mu_{\text{5 minutes of no intrument tasks}} < \mu_{\text{10 minutes of no instrument tasks}}$$

Notice that in this directional hypothesis, the researcher is asserting that with more non-instrument tasks comes greater achievement in vibrato performance. Although this may be a logical assumption to make, a researcher should be wary of directional hypotheses unless supported by evidence from previous researchers.

Setting Up a Database for a Kruskal Wallis H Test

The database for the Kruskal Wallis H test will have *two* necessary columns. One column will have the dependent variable (an ordinal or continuous variable) and the other column will have the independent variable (the grouping variable). In Figure 20.1, we can see a database established to compare the mean gain scores of four different groups of string players who have received different lengths of instruction using non-instrument tasks designed to improve vibrato (no instruction, 1 minute per class, 5 minutes per class, 10 minutes per class), players who have been assigned a number (i.e., no instruction = 1, 1 minute per class = 2, 5 minutes per class = 3, 10 minutes per class = 4) on their scores on a fictitious performance assessment (scaled 0–10).

Conducting a Kruskal Wallis H Test in SPSS

For this section, please use the provided example database, Kruskal Example 1 ⏵, in order to go through the process while you are reading. This exercise will give you the guided experience of a step-by-step analysis so that you can conduct the follow-up analysis on the additional dataset on your own.

In order to run the Kruskal Wallis H test, follow these steps:

1. Click **Analyze**.
2. Click **Nonparametric**.

	Group	VibratoMGSScore
1	1	3
2	1	4
3	1	3
4	1	4
5	1	3
6	1	2
7	1	3
8	1	4
9	1	3
10	1	4
11	1	3
12	1	3
13	1	4
14	2	3
15	2	4
16	2	2
17	2	3

FIGURE 20.1

3. Click **Legacy Dialog**.
4. Click **K Independent Samples**.
4. Enter the dependent variable in the **Test Variable** list. Note that SPSS will allow you to run multiple tests through this one screen by entering multiple dependent variables, but only one independent variable (factor).
5. Enter the independent variable in the **Grouping** space
6. Define range: enter the lowest number possible and the highest number possible. In this example the lowest numbered group is **1** and the highest numbered group is **4**.
7. Click **Continue**.
8. Make sure that you select **Kruskal Wallis H**.
9. In the **Options** screen make sure to select the **Descriptives** box.
10. Once you are back to the main window, click **OK**.

Once you have click **OK**, you should get three different boxes of information from SPSS: the descriptive statistics, the ranks, and the actual test statistics.

In the first box (see Figure 20.2), we see that the average mean gain score was 5 with a range of 2 to 9 and a standard deviation of 1.97. We also see that a total of 53 students participated in the study.

In the second box (Figure 20.3), we can see the ranking of the four groups, as well as how many participants were in each group. Keep in mind that the higher the mean rank, the higher the score on the vibrato test. So, we can see that group 4, those who received 10 minutes of non-instrument vibrato instruction, made the most gains, followed by group 3, then group 2, and, finally, group 1.

Descriptive Statistics

	N	Mean	Std. Deviation	Minimum	Maximum
VibratoMGSScore	53	5.00	1.971	2	9
Group	53	2.55	1.136	1	4

FIGURE 20.2

Ranks

	Group	N	Mean Rank
VibratoMGSScore	1	13	13.88
	2	12	14.25
	3	14	30.86
	4	14	46.25
	Total	53	

FIGURE 20.3

Test Statistics[a,b]

	VibratoMGSScore
Chi-Square	41.706
df	3
Asymp. Sig.	.000

[a]Kruskal Wallis Test
[b]Grouping Variable: Group

FIGURE 20.4

In the final box (Figure 20.4), we can see the actual test statistic, which in this case is a chi-square statistic, as well as the degrees of freedom and significance level. Here we see the chi-square statistic of 41.71 with three degrees of freedom (i.e., four groups at $N - 1$) and a significance of $p = .000$, or $p < .001$.

Do we have enough information to know where the difference is? Not yet. Although we know that at least one significant difference exists, we do not yet know between which groups the difference(s) exist. In order to examine where actual differences exist, we need to conduct some post hoc tests. One of the most direct, conservative, and easily interpretable means of doing this is through follow-up Mann-Whitney U tests. However, as this process means that you will be making multiple comparisons, it would be wise to use a Bonferroni adjustment (see Chapter 1) to mitigate a Type I error. This process is overly cautious, but will help the researcher get to the most interpretable findings. See Chapter 19 for how to run Mann-Whitney U tests.

After completing the Mann-Whitney U tests, we see that differences existed between groups 1 and 3 ($U = 13.5$, $p < .001$), groups 1 and 4 ($U = .00$, $p < .001$), groups 2 and 3 ($U = 13.0$, $p < .001$), and between groups 3 and 4 ($U = 3.5$, $p < .001$).

Is this enough information yet? Still not quite. We still have not talked about effect size. Although there is no clear way to calculate the effect size for the Kruskal Wallis H test itself, we can and should report the effect sizes of the follow-up Mann-Whitney U tests. See Chapter 19 for how to do this.

Reporting a Kruskal Wallis H Test

Once you have conducted the Kruskal Wallis H test and any post hoc tests by using the provided data set for this chapter, the final step is to create a research report that could be offered in a journal setting or dissertation writing. For readers not currently conducting their own research, looking at how these tests go from inception to reporting can give you a keen eye in determining the trustworthiness of the research report and the data contained within.

Given the data and analysis in our running music example in this chapter, one possible report of the data may look like Example 20.4.

Kruskal Wallis H Test Writing Hints

From the report in Example 20.4, we see some hints that might help you state your findings as clearly as possible while providing all of the necessary information.

- Make sure that you provide the following in your report:
  - All N or n data
  - Mean rank information
  - χ^2
  - df
  - p
  - U (as needed)
  - r (effect sizes for any post hocs)
  - Discussion of practical and statistical significance
- Per American Psychological Association guidelines, use first-person pronouns for clarity, especially when describing actions you took, while keeping all of your actions in past tense.
- Use the active voice.
- Avoid personification (e.g., "data indicated").
- Make sure all statistical indicators are in italics.
- Keep your actions as researcher in the past tense.

EXAMPLE 20.4 POSSIBLE KRUSKAL WALLIS H TEST REPORT OF THE EXAMPLE DATA SET

In order to examine the impact of different lengths of vibrato instruction without instruments in participants' hands, we employed a Kruskal Wallis H test. We elected to use this nonparametric test because of the relatively small number of participants in each group. Participants in group 1 ($n = 13$) received no vibrato instruction, while participants in group 2 ($n = 12$) received one minute of instruction per class. Those in group 3 ($n = 14$) received five minutes of vibrato instruction per class and those in group 4 ($n = 14$) received 10 minutes of vibrato instruction. We assessed participants' vibrato skill prior to treatment and again at the end of treatment. We used the mean gain score on these tests as the dependent variable and group as the independent variable. Participants in group 4 made the most gains with a mean rank of 46.25, followed by group 3 (mean rank = 30.86), group 2 (mean rank = 14.25), and group 1 (mean rank = 13.88). Using a Kruskal Wallis H Test, we found a significant difference between groups ($\chi^2 = 41.71$, $df = 3$, $p < .001$). In a post hoc analysis using Mann-Whitney U tests with a Bonferroni adjustment, we found the significant differences existed between groups 1 and 3 ($U = 13.5$, $p < .001$, $r = .75$), groups 1 and 4 ($U = .00$, $p < .001$, $r = .87$), groups 2 and 3 ($U = 13.0$, $p < .001$, $r = .74$), and between groups 3 and 4 ($U = 3.5$, $p < .001$, $r = .84$). On the basis of this data, we conclude that the differences between groups are both statistically as well as practically significant and warrant further exploration and consideration.

- Many readers will appreciate knowing the reason that you selected the nonparametric version of the test.

Kruskal Wallis H Test Exercises

Given the second provided dataset, Kruskal Wallis Example 2 ▶, compute the most logical analysis. In this example, the two columns are the following:

- Which elementary music class the third-grade students are in (1 = Ms. Smith's class, 2 = Mr. Jones' class, 3 = Ms. Walker's class)
- Score on composition project in which students composed a four-measure melody scored on a 5-point scale

In this example dataset, a researcher was interested in whether or not any differences existed among the three classes. Although each class had the same music teacher, the music teacher was curious if any differences existed among the classes. Following some anecdotal observation, the music teacher felt that students in Ms. Walker's class were not achieving at the same level.

Once you have computed the most appropriate statistical test, write a report that includes any descriptive statistics necessary to understand the Kruskal Wallis H test, any necessary post hoc tests, and any information needed to help the reader understand and trust your analysis.

21 | Spearman Correlation

Use of Spearman Correlation

As seen in Chapter 3, music education researchers can use correlation analysis in order to determine if a relationship exists between two variables and, if a relationship does exist, the relative magnitude of that relationship. The parametric version of this test (Pearson correlation) employs continuous data that is normally distributed. The nonparametric version of correlation to be discussed in this chapter, Spearman correlation (ρ or r_s), also known as Spearman's rho or Spearman's rank order correlation, can be used when the variables are ordinal or if the variables are continuous, but do not meet the assumptions of the Pearson correlation test. As with Pearson, the variables examined in a Spearman correlation do not have to be measured on the same scale. That is, you can examine the relationships between two different types of tests that are scored differently and on different scales. For example, a music education researcher may be interested in seeing if a relationship existed between middle-school choir students' GPA (4.0 scale) and an end-of-the-year music knowledge test scored on a 100-point scale.

It is important to remember when you are conducting this type of analysis that correlation *does not* show causation. As with the Pearson correlation, you cannot assume that because two things are related that one thing caused the other. Let's revisit the same analogy we made in the earlier chapter in regard to the Pearson correlation tests. Imagine, if you will, a music education professor asking graduate students in a large summer class to raise their hands if they feel that they have some skill in music. You could imagine that all the students in the room would raise their hands. Then the professor could look about the room and find that all the students are wearing some form of shoes. That would mean that there is a perfect direct correlation between wearing shoes and skill in music. On the basis of this assumption, you might conclude that wearing shoes leads to musical skill. Obviously, this is an extreme example, but it illustrates the logic of the process of correlation. Relationships may exist

> **EXAMPLE 21.1 USE OF SPEARMAN CORRELATION TO EXAMINE RELATIONSHIPS**
>
> It is also of interest to note that the perceived performance of *pedagogical content knowledge and skills* within the preservice course was positively correlated with early-career music teachers' overall satisfaction with the course (Spearman's rank order correlation—$\rho = .347$, $n = 74$, $p = .002$). This correlation further demonstrates the importance of *pedagogical content knowledge and skills* to early-career music teachers. Further research into this relationship could be helpful in reconceptualizing preservice programs.
>
> ---
>
> Ballantyne, J., & Packer, J. (2004). Effectiveness of preservice music teacher education programs: Perceptions of early-career music teachers. *Music Education Research*, 6(3), 299–312.

and it is important to learn about relationships. Causation, however, cannot be discussed on the basis of any correlational analysis.

With the use of a Spearman correlation, the relationship between two variables can range from −1.00 to 1.00. As with the Pearson correlation, a −1.00 indicates a perfect indirect correlation and a 1.00 indicates a perfect direct correlation. See Chapter 3 for a refresher regarding these relationships. The strength of the relationship functions the same as the Pearson correlation as well. The closer to −1.00 or 1.00 that a relationship gets, the stronger the relationship between the two variables. When you are discussing the strength of the relationship, the same guidelines for the Pearson correlation apply to the Spearman correlation. Correlations ranging from 0 to .29 (or −.29) are weak, .30 to .69 (or −.30 to −.69) are moderate, and anything larger than .70 (or −.70) is a strong correlation. It is with these measurements that a researcher can discuss the practical significance of a finding once statistical significance has been found.

As seen in Example 21.1, music education researchers can use Spearman correlation to examine relationships among many facets of music making and psychological or perceptual information. These researchers were interested in seeing if a relationship existed between music teaching skills as perceived in a class and overall satisfaction of the class.

Assumptions of Spearman Correlation

The Spearman correlation has several assumptions that are similar to those of other parametric tests. They include the following:

1. The variables need to be ordinal or continuous (ratio or interval).

2. Whereas in parametric correlation the relationship between the two variables should be linear, the relationship between any two variables being examined through Spearman correlation should be *monotonic*.

What does it mean to be monotonic? In a monotonic relationship, as one variable increases, the other either *only* increases or *only* decreases. In other words, it is much like the linear relationship of the parametric tests. See Figure 21.1 for examples of monotonic and non-monotonic relationships.

Although it is relatively rare to see an overt confirmation of the monotonic nature of a correlation, it does help the reader better understand the data being reported in a research report. See Example 21.2.

Research Design and Spearman Correlation

No research design really exists for the Spearman correlation. Rather, researchers use it to explore relationships between variables. Correlations are

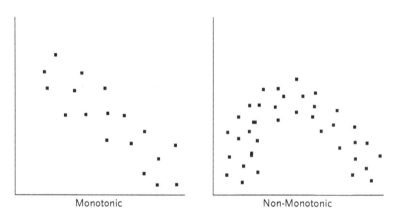

FIGURE 21.1

EXAMPLE 21.2 REPORTING OF MONOTONIC DATA IN SPEARMAN CORRELATION

Because both teacher ratings and student ratings suggested non-normality during examination of boxplots and skewness, a Spearman's rho correlation was calculated to determine the relationship between teacher rating and student rating values. There was a moderate, positive, monotonic correlation between teacher rating and student rating ($r_s = .595$, $n = 110$, $p < .001$, two-tailed).

Clementson, C. J. (2014). *A mixed methods investigation of flow experience in the middle level instrumental music classroom* (Unpublished Doctoral dissertation). University of Minnesota.

often used in descriptive studies and survey research studies. For example, a music education researcher may be interested to know if a relationship exists between high school choir students' perception of their choir director's conducting effectiveness and students' perceptions regarding the expressiveness of the final performance. In this case, the researcher could use a Spearman correlation (assuming the data did not meet the assumptions of a Pearson correlation) to see if indeed a relationship existed.

The Null Hypotheses for the Spearman Correlation

Although it is rare to overtly state the null hypothesis of the Spearman correlation, the null hypothesis (H_0) would be that no monotonic correlation exists between two variables. The alternative hypothesis (H_a) would be that a monotonic correlation does exist between two variables. Finally, a directional hypothesis (H_1) would indicate that a monotonic correlation exists and that the correlation is either direct or indirect, depending upon what we know from previous research.

Setting Up a Database for a Spearman Correlation

The database for a Spearman correlation requires at least two columns of variables between which the researcher wants to examine possible relationships. Both of these columns would contain data from two ordinal or continuous (interval or ratio) variables. As with the Pearson correlation, there really are no independent or dependent variables. In a bivariate (two variable) correlation

	ConductExpress	PerformExpress
1	4	10
2	5	9
3	4	8
4	5	9
5	4	8
6	3	9
7	4	8
8	5	8
9	5	7
10	4	8
11	5	9
12	4	8
13	4	7
14	4	8
15	5	9
16	5	8
17	5	7

FIGURE 21.2

analysis, the findings would be the exact same regardless of the order in which variables were analyzed. As we can see in Figure 21.2, this database has two columns. The first column contains students' mean score of a choir director's conducting expressiveness on a scale of 1 to 5. The second column contains the same students' rating of how expressive the final performance of the choir was on a scale of 1 to 10.

Conducting a Spearman Correlation in SPSS

For this section, please use the provided data set, Spearman Example 1 ▶ to conduct the analysis as you read. The first thing that the researcher must do is create a scatterplot of the data in order to see if the data is monotonic. Please refer to Chapter 3 (Pearson correlation) to see how to create scatterplots in SPSS. We can use the scatterplot to see if our data is monotonic or not. Once we are sufficiently convinced the data is monotonic, we can move forward with the analysis.

In order to run the Spearman Correlation in SPSS, follow these steps:

1. Click **Analyze**.
2. Click **Correlate**.
3. Click **Bivariate**.
4. *Move both variables over to the* **Variable List.** *Make sure to select* **Spearman,** *deselect* **Pearson,** *and select* **2-tailed Test.**
5. Click **Option**.
6. *Make sure that cases are excluded pairwise (this step keeps the largest number of participants possible in the analysis) by selecting* **Exclude Cases Pairwise.**
7. Click **Continue**.
8. Click **OK**.

Once you have clicked on **OK**, SPSS will provide you with one simple box. This box (Figure 21.3) provides all of the information from the Spearman correlation .

Correlations

			Conduct Express	Perform Express
Spearman's rho	Conduct Express	Correlation Coefficient	1.000	.400**
		sig. (2-tialled)	.	.000
		N	72	72
	Perform Express	Correlation Coefficient	.400**	1.000
		sig. (2-tialled)	.000	.
		N	72	72

**Correlation is significant at the 0.01 level (2-tailed).

FIGURE 21.3

From this box, we can see that 72 individuals participated in the study (or 72 people provided responses to both of the variables being analyzed). We can also see that the two variables are significantly associated in a directional relationship. We know that it is a direct relationship because the r_s is a positive number (i.e., .400) rather than a negative number (e.g., –.400). We can also learn from this box that this relationship is statistically significant ($p < .001$). On the basis of the guidelines discussed above, we can also identify this relationship as moderately strong (.40).

The matrix does repeat itself and compare variables to themselves. This repetition makes greater sense when you compare a larger number of variables, as it helps keep the matrix readable.

Unfortunately, what is not provided in this analysis is the descriptive data for these variables. In order to obtain the indicators of central tendency and variability, you will need to run descriptive analyses, as seen in Chapter 2.

Is this enough information? Yes. From the analysis thus far, you can report all of the descriptive statistics (after obtaining them through descriptive analysis, the r_s statistic, and the strength of the relationship).

Reporting a Spearman Correlation

Now that we have all the information that we need, we can begin to construct the research report of the Spearman correlation that could be given in a journal research report, thesis, or dissertation. Given the information from our analysis, one possible report of the data examined in the example in this chapter may look like Example 21.3.

Spearman Correlation Writing Hints

From the research report in Example 21.3, we see some hints that might help you state your findings as clearly as possible while providing all of the necessary information.

- Make sure that you provide the following in your report:
  - N
  - ρ or r_s
  - discussion of practical and statistical significance (in the case of correlation, it is the strength of the correlation)
  - means and standard deviations of all observations discussed
- Per American Psychological Association guidelines, use first-person pronouns for clarity, especially when describing actions you took, while keeping all of your actions in past tense.
- Use the active voice.
- Avoid personification.

> **EXAMPLE 21.3 POSSIBLE SPEARMAN CORRELATION REPORT OF THE EXAMPLE DATASET**
>
> High school choir students ($N = 72$) responded to two different items regarding expressivity in music. First, they rated the expressiveness of their director's conducting on a 5-point scale ($M = 3.88$, $SD = .99$). Secondly, the participants rated how expressive they felt their final choral performance was on a 10-point scale ($M = 6.93$, $SD = 1.41$). In order to see if these two phenomena related, I conducted a Spearman correlation analysis. I used this version of correlation because the data lacked a normal distribution for one of the variables. I found a statistically significant and moderate direct correlation ($\rho = .40$, $p < .001$) between students' perception of their director's expressivity while conducting and the expressivity of their final performance. The moderate magnitude of the relationship suggests that some practical significance to this finding may exist.

- Make sure all statistical indicators are in italics.
- Keep your actions as researcher in the past tense.
- Many readers will appreciate a quick discussion as to why you employed the nonparametric version of the test.

Spearman Correlation Exercises

Given the provided dataset, Spearman Correlation Example 2 ▶, compute the most logical analysis. In this example, a group of middle school brass players responded to a questionnaire asking about their comfort with three different brass techniques; lip slurs, flutter tonguing, and double tonguing. The columns are the following:

- Lip slur comfort on a scale of 1–10
- Flutter tonguing comfort on a scale of 1–10
- Double tonguing comfort on a scale of 1–10

Once you have computed the most appropriate statistical tests, write a report that includes all of the information required for a reader to assess the accuracy of your work, as well as the possible implications that may exist.

22 | Wilcoxon Test

Use of the Wilcoxon Test

As with the dependent samples t-test, researchers use the Wilcoxon test when they need to examine the differences in means between either the same group that has been tested twice (repeated-measures design) or two groups that have been paired or matched. If the researcher is examining the pre-test and post-test scores of the same group, each participant will need to have data collected at each point. If, however, the researcher is using a paired sample, then instead of having each individual's score for both tests, logical matching (i.e., similar backgrounds, experiences, age, etc.) is used to create pairs.

As seen in Table 1.6 in Chapter 1, researchers using the Wilcoxon test are trying to answer a simple research question: does a difference exist between two observations of the same phenomenon (paired samples) on one dependent variable?

For example, if the researcher wants to see if a change has occurred between a pre-test and a post-test of some skill and the data does not meet the assumptions of the parametric dependent-samples t-test, the Wilcoxon test would be appropriate. A music education researcher may be interested in whether or not students made significant gains in their ability to improvise over rhythm changes. Students could be given a pre-test at the start of the improvisation unit and then the same students could take a post-test at the end of the instructional unit to see if they had made a significant advancement.

When researchers use the Wilcoxon test to examine differences between the means of the same group over a period of time, the dependent variable is usually derived from a research instrument that is used twice (either in the same form or in a parallel form that has been shown to be reliable with the first form of the test). Generally, some sort of treatment is received between the pre-test and post-test. In our example, it would be the improvisation instruction. The same is true for a paired sample except that the researcher did not collect data twice from the same participants.

> ### EXAMPLE 22.1 WILCOXON TEST REPORT
>
> Responses on the Rosenberg Self-Esteem Scale (SES), the Index of Peer Relations (parent and participant version) (IPR), and the State-Trait Anxiety Inventory—trait version (STAI) were compared pre- and post-program intervention in a within-groups design. Data from both groups of the music program was combined. Data was analyzed nonparametrically because of the type of data and small participant numbers (Wilcoxon signed ranks test), and hypotheses were two-tailed.
>
> Data from the weekly STAI-state version was combined across weeks 2–7. Self-report ratings of anxiety were significantly lower at the end of the music sessions compared to ratings given at the beginning of the sessions ($n = 125$; Wilcoxon: $Z = 5.077$; $p = 0.001$).
>
> ---
>
> Hillier, A. J., Greher, G., Poto, N., & Dougherty, M. (2011). Positive outcomes following participation in a music intervention for adolescents and young adults on the autism spectrum. *Psychology of Music*, 0305735610386837.

In Example 22.1, music researchers used the Wilcoxon test to see if significant change occurred in adolescents' anxiety between a pre-test and post-test. Between the two tests, participants engaged in a music seminar.

Assumptions of the Wilcoxon Test

The Wilcoxon test really has only two major assumptions that will be taken care of on the basis of the design of the study and the type of data collected:

1. The data collected for the dependent variable (e.g., pre-test and post-test scores) is ordinal or continuous rather than categorical.
2. Each observation is independent of any other observation (most often accomplished through random sampling). Don't confuse this assumption with having scores from the same person. In a dependent-samples t-test the score of the pre-test and post-test are most likely related, but the score of one participant is not dependent upon the score of another participant.

Again, normally we would need to argue for normal distribution of the data as well. However, as the Wilcoxon test is nonparametric, the researcher does not need to be concerned about normality.

Meeting the Assumptions of a Wilcoxon Test

As discussed above, meeting the assumptions of this test is really a matter of the design of the study and the type of data collected. As long as the dependent variable is at least ordinal and there are no more than two observations of the dependent variable, this test should be relatively robust against most violations of any assumptions.

Research Design and Wilcoxon Test

The Wilcoxon test can be used to compare the change over time or between the treatment of any group. So, the possible research designs that utilize this type of inferential test is primarily (O = observation at ordinal or continuous level, X = some treatment):

(A) O   X   O

If the research uses this design, the Wilcoxon test is going to help us identify any change in mean between the pre-test and post-test for the one group included in the design.

Let's take a look at what this design might look like with a musical example.

(A) $O_{\text{improvisation skill pretest}}$ $X_{\text{improvisation instruction}}$ $O_{\text{improvisation skill posttest}}$

In this example, the researcher was interested to see if her planned improvisation instruction actually helped students develop their ability to improvise over rhythm changes. To achieve this, she created and administered an improvisation pre-test, taught the unit on improvisation, and then collected data at the end of instruction on students' ability to improvise. However, because of a small class size, her data did not meet the assumptions of the dependent-samples t-test; therefore, she was correct in using the Wilcoxon test.

The Null Hypothesis for a Wilcoxon Test

The null hypothesis in a Wilcoxon test is the same as the dependent-samples t-test: that the mean of the pre-test will not be different from the mean of the post-test. It could be stated as

$H_0: \mu_{\text{pretest}} = \mu_{\text{posttest}}$

For our musical example regarding improvisation skill development, the null hypothesis could be stated as

$H_0: \mu_{\text{improvisation pretest}} = \mu_{\text{improvisation posttest}}$

The alternative hypothesis would be that there would be a difference between the mean of the pre-test and post-test scores. It could be stated as

$H_a: \mu_{\text{pretest}} \neq \mu_{\text{posttest}}$

In our musical example, the alternative hypothesis could be labeled as:

$H_a: \mu_{improvisation\,pretest} \neq \mu_{improvisation\,posttest}$

Finally, the directional hypothesis would state that the mean of the pre-test and post-test would not be equal and that one would be higher or lower than the other. The most common directional hypothesis used with a Wilcoxon test is that the post-test mean will be higher than the pre-test. Even though this is the logical outcome of a research design such as our example, a directional hypothesis should not be made unless the researcher had evidence from previous research that it is likely to happen. This hypothesis would be stated as

$H_1 = \mu_{prestest} < \mu_{posttest}$

In our musical example, it would be stated as

$H_1: \mu_{improvisation\,prestest} < \mu_{improvisation\,posttest}$

Setting Up a Database for a Wilcoxon Test

The database for a Wilcoxon test will have two columns. In one column, you would find the data from the first observation of the participants (e.g., the improvisation pre-test). In the second column, you would include the data from the second observation of the participants (e.g., the improvisation post-test), as can be seen in Figure 22.1. In Figure 22.1, we can see that the scores for the theoretical improvisation pre-test (possible range of 0–10) are in the first column while the theoretical scores for the post-test (on the same scale) can be found in the second column.

	ImprovPRE	ImprovPOST
1	8	10
2	9	10
3	8	10
4	7	10
5	8	10
6	8	10
7	8	10
8	7	10
9	7	10
10	8	10
11	6	10
12	5	7
13	7	8
14	6	8
15	6	7

FIGURE 22.1

Conducting a Wilcoxon Test in SPSS

For this section, please use the provided example dataset, Wilcoxon Example 1 ▶ as you read so that you can gain experience in conducting this test while learning about it. In order to conduct the Wilcoxon test in SPSS, follow these steps:

1. Click **Analyze**.
2. Click **Nonparametric**.
3. Click *Legacy* **Dialog**.
4. Click **2 related samples**.
5. Move the pre-test variable over to the **First Pair Test Variable** *slot*.
6. Move the post-test variable into the **Second Test Variable** *box*.
7. If you have multiple comparisons to run, you can enter more than one pair of variables at a time.
8. Click **Options**.
9. Click **Descriptives**.
10. Click **Continue**.
11. *Make sure to select* **Wilcoxon**.
12. Click **OK.**

Once you have clicked **OK**, you should see a report with three boxes of information. The first box (Figure 22.2) gives you the descriptive statistics that inform the following analyses. In this box, we can see the number of participants in the study (in this case, 46), the mean score on the pre-test (5.83) and the mean score on the post-test (8.13) and both scores' standard deviations and ranges.

In the second box (Figure 22.3), we can see the ranks of the scores as well as the directions of the ranks. More specifically, in Figure 22.3 we can see that 1 student scored lower on the post-test than on the pre-test (negative rank), 42 students scored higher on the post-test than the pre-test, and 3 students scored the same on the pre-test and post-test.

The final box provides the actual outcome of the Wilcoxon test, which is a Z statistic, as well as the statistical significance of the test. As seen in Figure 22.4, the Wilcoxon Z statistic is –5.575, which is significant at the .000, or $p < .001$.

Is this enough information? Not yet. Although many researchers stop at reporting the statistical significance of the Wilcoxon test, more information would be most helpful to determine the practical significance of the finding.

Descriptive Statistics

	N	Mean	Std.Deviation	Minimum	Maximum
ImprovPRE	46	5.83	1.924	2	9
ImprovPOST	46	8.13	1.439	6	10

FIGURE 22.2

Ranks

		N	Mean Rank	Sum of Ranks
Improv POST - Improv PRE	Negative Ranks	1[a]	16.00	16.00
	Positive Ranks	42[b]	22.14	930.00
	Ties	3[c]		
	Total	46		

[a] Improv POST < Improv PRE
[b] Improv POST > Improv PRE
[c] Improv POST = Improv PRE

FIGURE 22.3

Test Statistics[a]

	ImprovPoST − ImprovPRE
Z	−5.575[b]
Asymp. sig. (2-tailed)	.000

[a] Wilcoxon Signed Ranks Test
[b] Based on negative ranks.

FIGURE 22.4

As with the Mann-Whitney test, SPSS does not create a report of effect size automatically. However, computing the effect size (r) for this test is quite simple and looks very similar to that found for the Mann-Whitney U test. This is a straightforward formula that will take you only a few moments to compute, but gives your analysis much greater depth and meaning. The formula for r (effect size for the Wilcoxon test) is

$$r = \frac{Z}{\sqrt{N_1 + N_2}}$$

In this formula, we obtain Z from the third box provided in the SPSS output (see Figure 22.4) and we know that the N of the first observation is 46 and the N of the second observation is 46. With this information, we can begin to calculate r.

$$r = \frac{-5.575}{\sqrt{46 + 46}}$$

$$r = \frac{-5.575}{\sqrt{92}}$$

$$r = \frac{-5.575}{9.59}$$

$$r = -.58$$

EXAMPLE 22.2 POSSIBLE WILCOXON TEST RESEARCH REPORT OF THE EXAMPLE DATA

In order to determine if high school students participating in a jazz ensemble made significant gains in improvisation skills following a unit of improvisation instruction, I employed a Wilcoxon test to determine any differences between pre-test scores and post-test scores. I employed the nonparametric Wilcoxon test because of a small class size. Participants in the study ($N = 46$) scored an average of 5.83 ($SD = 1.92$) on the pre-test and 8.13 ($SD = 1.44$) on the post-test. Through my analysis, I found that this gain was statistically significant (Wilcoxon $Z = -5.575$, $p < .001$). In order to explore the practical significance of this finding, I computed the effect size and found a large effect size ($r = .58$), suggesting practical as well as statistical significance.

In general, Wilcoxon tests (as with Mann-Whitney U tests) with an r of .1 to 2.9 are small effect sizes, .3 to .49 are moderate effect sizes, and .5 and above are large effect sizes. So, in our example, we have a rather large effect size and can therefore argue for practical significance of our findings.

Reporting a Wilcoxon Test

Once you have conducted the Wilcoxon test and computed the effect size, you have enough information to write a full report of your findings. Given the information we found in our running example, one possible report of the data may look like the one found in Example 22.2.

Wilcoxon Writing Hints

From the research report in Example 22.2, we see some hints that might help you state your findings as clearly as possible.

- Make sure to report the following:
  - Means
  - Standard deviations
  - Wilcoxon Statistic (Z)
  - The significance
  - The effect size (r)
    - Do this for all of the analyses you conduct even if they are not statistically significant.
- Per American Psychological Association guidelines, use first-person pronouns for clarity, especially when describing actions you took, while keeping all of your actions in past tense.

- Keep your reporting in past tense.
- State your findings in the active voice.
- Remember to avoid personification. Results and studies do not "do" anything.
- Remember to place the statistical identifiers (e.g., t, SD, M) in italics. Do not forget to report the means and standard deviations of the descriptive data.
- Many readers will appreciate a quick discussion as to why you employed the nonparametric version of the test.

Wilcoxon Test Exercises

Given the second provided dataset, Wilcoxon Example 2 ⊙, compute the most logical analysis. In this example, the columns include the following:

- Pre-test and post-test for a pitch-matching test each with a scale of 0 to 10.

In this study, the researcher was interested in finding out if his elementary students improved their ability to match pitch after having musical instruction for nine weeks. Because of the small number of students in his early elementary classes, he felt it most prudent to use the nonparametric Wilcoxon test.

Conduct an analysis of the paired samples. Once you have computed the most appropriate statistical test, write a report that includes everything the reader needs to know to understand and trust your analysis.

23 | Friedman's Test

Use of Friedman's Test

Friedman's test is the nonparametric equivalent to a one-way repeated-measures ANOVA. As such, the test can be used by music education researchers when the same groups of individuals have been observed at three or more different times. If only two observations have been made, it would be more appropriate use a dependent-samples t-test if the data was normally distributed or a Wilcoxon test if the nonparametric test was required. As with the repeated-measures ANVOA, the Friedman's test allows a music education researcher to examine change over time within a single group of participants.

As seen in Table 1.6 in Chapter 1, researchers using a Friedman's test are trying to answer one question: does a difference exist between three or more observations of the same phenomenon (paired sample) on one dependent variable?

For example, a music education researcher may be interested to see if an instructional unit focused on tonal analysis had a long-term impact on the score-reading ability of undergraduate music education majors. The researcher could evaluate students' score-reading ability once or twice prior to the unit of instruction along with a couple of times after the unit to see if any improvement was due to simple maturation or the treatment, as well as to see if the treatment had long-term impact on students' score reading ability.

As with a one-way repeated-measures ANOVA, the Friedman's test requires one ordinal or continuous dependent variable and one categorical independent variable. However, the independent variable in a Friedman's test is usually the different data-collection points themselves. For example, our music researcher who was interested in the impact of tonal-analysis instruction of score-reading skill tested participants' skills twice before the treatment and twice after the treatment. In this example, the dependent variable is the scores students received on their score-reading test while the independent variable is the number of observation: that is, observation 1 compared with observation 2, and the rest of the series.

> ### EXAMPLE 23.1 USING A FRIEDMAN'S TEST TO EXPLORE CHANGES TO RESPONSES TO THE SAME QUESTIONS WITH DIFFERENT TREATMENTS BETWEEN OBSERVATIONS
>
> The daily-groups means for each of the four questions on the survey (I like this music and I like rehearsing this music—for both selections) were compared across the four treatment days by using a Friedman test, and the results were statistically significant for both the beginning group—$\chi_r^2 (4, 4) = 9.9, p < .05$—and the advanced group—$\chi_r^2 (4, 4) = 12, p < .05$). Day 1 was the lowest ranked, followed by day 2, then day 3. Day 4 was the highest ranked. This finding indicates that the least positive attitudes were held on the first day the music was presented, when the participants were sight-reading unfamiliar music, and the most positive attitudes were held on the last day of the experiment, when they were most familiar with the music.
>
> ---
>
> Napoles, J. (2007). The effect of duration of teacher talk on the attitude, attentiveness, and performance achievement of high school choral students. *Research Perspectives in Music Education, 11*(1), 22–29.

In Example 23.1, a researcher used a Friedman's test to see if participants' responses to the same questions regarding the music they were working with that day changed over the course of four days.

Assumptions of Friedman's Test

The nonparametric Friedman's test has only two assumptions and, as with most nonparametric tests, they are usually decided upon during the study design phase. The assumptions of the Friedman's test are the following:

1. One group of participants are observed or measured three or more times.
1. The data collected for the dependent variable (e.g., the outcome of the score reading test) is either ordinal or continuous.

Research Design and Friedman's Test

The research designs for the Friedman's test are the same as the research designs for the repeated-measures ANOVA and can be used in any research design in which the researcher has collected data that is either ordinal or non-normally distributed continuous data from one group multiple times.

(A) O —— O —— O

(B) O —— O —— O —— X —— O —— O —— O

(C) O —— X —— O —— X —— O —— X —— O

Or the test can be used for any variation of a time-series design of one group with any given observations and treatments.

Let's take a look at a few musical examples of what these designs may look like in the field of music education.

(A) $O_{\text{hearing test}}$ —— $O_{\text{hearing test}}$ —— $O_{\text{hearing test}}$ —— etc.

In our first musical example (A), a researcher may be interested in the loss of hearing in a group of music teachers. In order to explore the hearing loss (or not) of music teachers, you might use this design, which has no specific treatment. Actually, the researcher is interested in the impact of time on music teachers' hearing. The researcher is not controlling for such factors as the acoustics of the room, size of ensemble, type of ensemble, and number of minutes with each day. The only conclusion that you would be able to draw by using this design is that music teachers' hearing is or is not changing significantly over time, a conclusion that might then encourage and inform the next series of studies to be undertaken.

(B) $O_{\text{score reading test}}$ — $O_{\text{score reading test}}$ — $X_{\text{tonal analysis unit}}$ — $O_{\text{score reading test}}$ — $O_{\text{score reading test}}$

In musical example B, we see a study we have already discussed. A researcher was interested to see if a unit of instruction on tonal analysis had a significant impact on students' score-reading ability. This researcher tested the students' score-reading skills twice before the treatment and twice following the treatment. This design has a few benefits. If students improve, but improve at the same rate between all observations, we might conclude that the treatment was not the thing that made the impact. We might find a significant bump in the third observation, indicating some impact of the treatment. That impact might either continue or scores might regress toward the mean in the fourth observation, which will let us know whether or not the treatment seems to have a long-term effect. In this design, the dependent variable is the score on the four score-reading tests, the independent variable is the observation (i.e., observation 1, 2, 3, or 4).

(C) O_{enjoy} — $X_{\text{opus 1}}$ — O_{enjoy} — $X_{\text{opus 2}}$ — O_{enjoy} — $X_{\text{opus 3}}$ — O_{enjoy}

Our last musical example (C) looks much like the one found in Example 23.1. In this example, participants rated the extent to which they enjoyed rehearsal prior to and after sight-reading three new works. This researcher may be interested in seeing how the type of music selected impacts how much students enjoy the rehearsal. If the ensemble director believes that student enjoyment is an important facet of a program, this information may help her make informed decisions about what types of repertoire to include in the future.

The Null Hypotheses for Friedman's Test

Researchers can use a Friedman's test to test the null hypothesis that mean ranks for the group examined will not be different at different observations. For example, it could be stated as

$H_0: \mu_1 = \mu_2 = \mu_3 = \mu_4 = \mu_5$ etc.

For our musical example above regarding the enjoyment of a musical rehearsal, the null hypothesis would be that there is no difference in enjoyment between any of the days regardless of what type of piece is rehearsed. It could be stated as

$H_0: \mu_{enjoy} = \mu_{enjoy} = \mu_{enjoy} = \mu_{enjoy} = \mu_{enjoy}$ etc.

The alternative hypotheses for the Friedman's test would be that a difference did exist between observations and could be stated as

$H_a: \mu_1 \neq \mu_2 \neq \mu_3 \neq \mu_4 \neq \mu_5$ etc.

In our musical example, the alternative hypothesis would be that students did enjoy the rehearsals differently and it could be stated as

$H_a: \mu_{enjoy} \neq \mu_{enjoy} \neq \mu_{enjoy} \neq \mu_{enjoy} \neq \mu_{enjoy}$ etc.

Finally, the directional hypothesis for a Friedman's test is that between each observation outcomes will be different in a specific direction. This could be stated as

$H_1: \mu_1 < \mu_2 < \mu_3 < \mu_4 < \mu_5$ etc.

If, in our musical example, the researcher had reason to believe that she had placed the works in an order that was most likely to move from least enjoyable to most enjoyable, then the directional hypothesis might be stated as

$H_a: \mu_{enjoy} < \mu_{enjoy} < \mu_{enjoy} < \mu_{enjoy} < \mu_{enjoy}$ etc.

Setting Up a Database for a Friedman's Test

The database for a Friedman's test looks just like that of a repeated-measures ANOVA and should have at least three columns. If you had only two columns, it would most likely be more appropriate to conduct a Wilcoxon test. The three or more columns should contain information from the same participants (each row should have data from one individual), while each column should represent a different condition or time that the participants was observed or measured. In Figure 23.1 we see an example of the database for our musical example in which we have four columns, the two score-reading tests prior to the tonal-analysis unit and the two score-reading tests following the tonal-analysis unit.

Conducting a Friedman's Test in SPSS

For this next section, please use the provided dataset, Friedmans Example 1 ▶, to conduct the analysis as you read. Conducting a Friedman's test in SPSS is relatively simple and can be completed by following these steps:

1. Click **Analyze**.
2. Click **Nonparametric tests**.
3. Click **Legacy Dialog**.
4. Click **K related samples**.
5. Enter all of the observations into the **Test variables** slot.
6. Click **Statistics**.
7. Select **Descriptives**.
8. Click **Continue**.

	ScoreReading1	ScoreReading2	ScoreReading3	ScoreReading4
1	11	12	14	14
2	13	13	16	15
3	14	14	17	17
4	12	12	16	16
5	13	13	17	17
6	12	13	16	17
7	13	13	17	16
8	14	13	15	18
9	15	16	18	17
10	16	17	19	18
11	13	13	17	17
12	13	12	17	16
13	9	10	15	17
14	9	11	14	15
15	8	10	16	16

FIGURE 23.1

9. *Make sure to select* **Friedman**.
10. *Click* **OK**.

Once you have clicked **OK**, the program will produce three boxes of data. The first box (Figure 23.2) provides the descriptive statistics for the analysis. As we can see in Figure 23.2, 20 students participated in this study. The mean score for the first observation was 11.95 with a standard deviation of 2.19. The mean score for the second observation was 12.40 with a standard deviation of 1.82. The mean score for the third observation was 15.85 with a standard deviation of 1.46, while the mean for the final observation was 16.10 with a standard deviation of 1.17.

In the second box (Figure 23.3), we can see the mean ranks of the four observations. From Figure 23.3, we can see that the mean rank of the first score-reading test was 1.33. The mean ranks increased from there for the following three observations at 1.68, 3.45, and 3.55, respectively.

In the final box created in SPSS for this test, we can see the actual test statistic, which in this case is a chi-square outcome (see Figure 23.4). We can see that the chi-square statistic for this analysis is 52.75 with three degrees of

Descriptive Statistics

	N	Mean	Std. Deviation	Minimum	Maximum
Score Reading 1	20	11.95	2.188	8	16
Score Reading 2	20	12.40	1.818	10	17
Score Reading 3	20	15.85	1.461	14	19
Score Reading 4	20	16.10	1.165	14	18

FIGURE 23.2

Ranks

	Mean Rank
Score Reading 1	1.33
Score Reading 2	1.68
Score Reading 3	3.45
Score Reading 4	3.55

FIGURE 23.3

Test Statistics[a]

N	20
Chi-Square	52.751
df	3
Asymp. Sig.	.000

[a]Friedman Test

FIGURE 23.4

freedom and a significance level of $p < .001$. So, in this case, we have a significant outcome. We do not know, however, where the significance exists. Is it between all four observations or just between select ones? We need to do some further analysis.

One of the best ways to examine the outcome of any repeated measures design is through a graph. In order to create a graph of these findings, follow these steps:

1. Click **Graphs**.
2. Click **Legacy Dialogs**.
3. Click **Line**.
4. Select **Simple line**.
5. Before clicking **Define**, *make sure to select* **Summaries of separate variables**.
6. Click **Define**.
7. Enter each of the variables in order or observation (i.e., **observation 1** *first,* **observation 2** *second, etc.*).
8. Click **OK**.

Once you have clicked **OK**, SPSS will create a graph of the data that you can use to visually inspect where you think the significant differences may have occurred (see Figure 23.5).

Although visually we might conclude that the significant difference would be between observation 2 and observation 3, we need to verify that outcome with more analysis. In order to run the post hoc tests needed to examine where the differences exist, we will need to complete a serious of Wilcoxon tests (see Chapter 22) with a Bonferroni adjustment to help mitigate Type I error.

After completing the Wilcoxon tests, we see that the only significant difference existed between observation 2 and observation 3 with a large effect size (Wilcoxon $Z = -.95$, $p < .001$, $r = .63$), which is when we had our treatment as our research design was

$$O_{\text{score reading test}} - O_{\text{score reading test}} - X_{\text{tonal analysis unit}} - O_{\text{score reading test}} - O_{\text{score reading test}}$$

Now we have enough information to report the findings of our Friedman's test analysis.

Reporting a Friedman's Test

Once you have conducted the Friedman's test and have all of the follow-up information you need, the next step is to write a report of the data that will give your readers all the information they need to understand and trust your analysis.

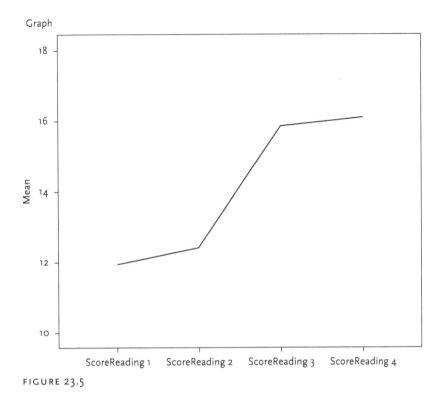

FIGURE 23.5

Given the outcomes from the analysis of our musical example in this chapter, one possible report of the data may look like the one found in Example 23.2.

Friedman's Test Writing Hints

From the research report in Example 23.2, we see some hints that might help you state your findings as clearly as possible while providing all of the necessary information.

- Make sure that you provide the following in your report:
    - N
    - Means and standard deviations of all observations discussed
    - Mean ranks
    - χ^2
    - df
    - p
- Discussion of practical and statistical significance
- Post hoc analysis (including Z, p, and r)
- Per American Psychological Association guidelines, use first-person pronouns for clarity, especially when describing actions you took, while keeping all of your actions in past tense.

EXAMPLE 23.2 POSSIBLE FRIEDMAN'S TEST REPORT OF THE EXAMPLE DATASET

In order to examine the impact of an instructional unit focused on tonal analysis on a class of undergraduate music education students' score-reading skill, I evaluated participants' ($N = 20$) score-reading skills twice before teaching the unit of tonal analysis and twice after analysis. I used this design because I had access to only one class of students and because I hoped to be able to eliminate the threat of maturation. Participants did score higher on each successive score-reading evaluation. The mean score for the first observation was 11.95 with a standard deviation of 2.19. The mean score for the second observation was 12.40 with a standard deviation of 1.82. The mean score for the third observation was 15.85 with a standard deviation of 1.46, while the mean for the final observation was 16.10 with a standard deviation of 1.17. The mean ranks of each successive evaluation increased as well. The mean rank of the first score-reading test was 1.33. The mean ranks increased from there for the following three observations at 1.68, 3.45, and 3.55, respectively. I employed a Friedman's test to see if these gains were significant and found a significant difference between observations ($\chi^2 = 52.75$, $df = 3$, $p < .001$). In order to see where the differences existed, I completed a series of Wilcoxon tests with a Bonferroni adjustment and found that the only significant difference between the second and third observation (Wilcoxon $Z = -3.95$, $p < .001$, $r = .63$). I found that this difference had a large effect size, suggesting that instruction in tonal analysis is a beneficial practice when instructors are hoping to improve the score-reading skills of pre-service music educators.

- Use the active voice.
- Avoid personification.
- Make sure all statistical indicators are in italics.
- Keep your actions as researcher in the past tense.
- Many readers will appreciate a quick discussion as to why you employed the nonparametric version of the test.
- Although not included here, the graph that displays the difference between the groups can be a very helpful visual aid for a reader.

Friedman's Test Exercises

Given the provided dataset, Friedman's Example 2 ⓑ, compute the most logical analysis. In this example, the columns are scores from middle school orchestra students who rated their enjoyment of rehearsals on the basis of the new music they worked on that day:

- Enjoyment of rehearsal prior to getting new music
- Enjoyment of rehearsal after new music 1

- Enjoyment of rehearsal after new music 2
- Enjoyment of rehearsal after new music 3
- Enjoyment of rehearsal after new music 4

The research design for this study would look thus:

O —— X —— O —— X —— O —— X —— O —— X—— O

This analysis might have you thinking like a teacher who is concerned about how our repertoire selection may impact student enjoyment and then possibly retention and achievement. If you find any significant changes, examine the plot to see where the changes may exist and then complete any needed post hoc tests. Begin to think about the implications of these findings and how you might explain them in a discussion section.

Once you have computed the most appropriate statistical tests, write a report that includes any descriptive statistics necessary to understand and trust the Friedman's test outcome and your analysis.

APPENDIX

IN THIS APPENDIX, THE READER can find the majority of statistical outcomes necessary to complete the second writing exercises in each of the chapters found in this text. As this appendix is intended as a guide to make sure the reader is on the right track rather than provide all of the answers and data required, not all information that should be reported might be found. Descriptive statistics, for example, might be omitted in this appendix unless they are germane to the nature of the analysis. Additionally, some post hoc outcomes are not overtly listed, although the type of post hoc test is stated to help the reader complete the correct analyses.

Chapter 2: Descriptive Statistics

$N = 30$
Gender = 17 male (56.7%), 13 female (43.3%)
Teaching Genre = 6 orchestra teachers (20%), 20 choir teachers (40%), 20 band teachers (40%)
Teaching Experience range = 32
Teaching Experience mean = 11.73
Teaching Experience $SD = 10.62$
Teaching Evaluation range = 4
Teaching Evaluation mean = 2.63
Teaching Evaluation $SD = 1.30$
Student Average Score range = 30
Student Average Score mean = 90.5
Student Average Score $SD = 8.03$
Average number of students in all-state range = 9
Average number of students in all-state mean = 3.70
Average number of students in all-state $SD = 2.89$

Chapter 3: Pearson Product Moment Correlation

Significant relationships:
Teaching Effectiveness—Teaching Experience ($r = .87, p < .001$)

Teaching Effectiveness—Teaching Joy ($r = -.87, p < .001$)
Teaching Experience—Teaching Joy ($r = -.82, p < .001$)

Chapter 4: One-Sample T-Test

$N = 74$
Mean = 7.59
$SD = 1.35$
$t = -.670$
$df = 73$
significance = .505
mean difference = $-.105$

Chapter 5: Dependent-Samples T-Test

Significant outcome:
PretestSR—PosttestSR ($t = -15.48, df = 49, p < .001$)
Not significant outcome:
PretestMA—PosttestMA ($t = 1.48, df = 49, p = .15$)

Chapter 6: Independent-Samples T-Test

Levene outcome: $F = .002, p = .97$
t-test outcome: $t = -19.80, df = 78, p < .001$

Chapter 7: Univariate ANOVA

Levene outcome: $F = 2.17, p = .09$
ANOVA outcome: $F = 7.56, df = 3, p < .001$
On the basis of post hoc tests (Scheffe), differences existed between group 1 and all of the other groups.

Chapter 8: Factorial ANOVA

Levene outcome: $F = 1.24, p = .30$
Factorial ANOVA outcome:
An interaction existed between treatment and instrument voice: $F = 101.48, df = 1, p < .001$, partial $\eta^2 = .54$).

Chapter 9: MANOVA

Box's M outcome: Box's M = 13.03, $F = 1.05, p = .40$
Wilks's lambda outcome: $\Lambda = .011, F = 251.54, df = 8, p < .001$, partial $\eta^2 = .89$)
MAP Levene outcome: $F = 2.65, p = .04$
Sightsing Levene outcome: $F = .33, p = .86$
MAP ANOVA outcome: $F = 1241.41, df = 4, p < .001$, partial $\eta^2 = .98$
Sightsing ANOVA outcome: $F = 55.69, df = 4, p < .001$, partial $\eta^2 = .65$
MAP post hoc (Games-Howell) show significant differences between groups:
1/2 1/4 2/4 2/3 2/4 2/5 3/4 4/5 Sightsing post hoc (Scheffe) show significant differences between groups:
1/2 1/3 1/4 1/5 2/3 2/4 3/5 4/5

Chapter 10: Repeated-Measures ANOVA

Mauchly's test of sphericity outcome: Mauchly = .91, $\chi^2 = 6.87, p = .65$

Repeated-measures ANOVA outcome: $F = 89.74$, $df = 4$, $p < .001$, partial $\eta^2 = .55$
Post hoc tests show significant differences between observations 3 and 4 and observations 4 and 5.

Chapter 11: ANCOVA

Levene outcome: $F = 1.00$, $p = .37$)
ANCOVA outcome $F = 11.96$, $df = 1$, $p = .001$, partial $\eta^2 = .12$
Adjusted mean group 1 = 4.60, standard error = .09
Adjusted mean group 2 = 4.87, standard error = .07
Adjusted mean group 3 = 3.93, standard error = .09
Post hoc tests demonstrate differences between Group 3 and the other two groups.

Chapter 12—MANCOVA

Box's M outcome: Box's M = 8.18, $F = .88$, $p = .54$
Grade Wilks's lambda outcome: ($\Lambda = .998$, $F = .16$, $df = 2$, $p = .86$, partial $\eta^2 = .002$)
Ensemble Wilks's lambda outcome: ($\Lambda = .39$, $F = 26.07$, $df = 6$, $p < .001$, partial $\eta^2 = .38$)
Theory Levene outcome: $F = .46$, $p = .71$
Interest Levene outcome: $F = 1.03$, $p = .38$
Ensemble:
Theory ANOVA outcome: $F = 29.01$, $df = 3$, $p < .001$, partial $\eta^2 = .40$)
Interest ANOVA outcome: $F = 27.79$, $df = 3$, $p < .001$, partial $\eta^2 = .39$)
Post hoc tests indicate differences between the following groups (ensembles) for theory:
1/4 2/3 2/4 3/4 Post hoc tests indicate differences between the following groups (ensembles) for interest:
1/2 1/3 1/4 2/4 3/4

Chapter 13: Regression Analysis

$R = .84$
$R^2 = .70$
Adjusted $R^2 = .69$
ANOVA outcome: $F = 56.21$, $df = 3$, $p < .001$
Significant predictors:
Rhythm ($\beta = .14$, $t = 2.09$, $p = .04$, tolerance = .96, VIF = 1.05)
Harmony ($\beta = .81$, $t = 11.74$, $p < .001$, tolerance = .88, VIF = 1.14)

Chapter 14: Factor Analysis

If the researcher used a PCA with a promax rotation:
KMO = .836
Bartlett's test of sphericity ($\chi^2 = 4874.24$, $df = 946$, $p < .001$)
Variance explained = 67.82%
Number of components: 12 (to be explained and labeled by the researcher)

Chapter 15: Discriminant Analysis

Following bivariate analyses, the only variables to be included are the following:
Intonation

Style
Tempo
Phrasing
Interpretation
Box's M outcome: Box's M = 36.82, $F = 1.14$, $p = .28$
Function 1 eigenvalue = .30
Function 2 eigenvalue = .13
Wilks's lambda outcome: ($\Lambda = .68$, $\chi^2 = 37.06$, $df = 10$, $p < .001$)
41.4% of group 1 are correctly classified
73.5% of group 2 are correctly classified
62.2% of group 3 are correctly classified
60% overall are correctly classified.

Chapter 16: Cronbach's Alpha

Alpha = .29
Tukey: $F = 1.97$, $p = .16$

Chapter 17: Split-half Reliability

Goodness of fit: ($\chi^2 = 1131.37$, $df = 89$, $p < .001$)
Guttman coefficient = .79
Tukey: $F = 17.10$, $p < .001$

Chapter 18: Chi Square

$\chi^2 = 1.023$, $df = 1$, $p = .322$
Phi = $-.069$, $p = .31$

Chapter 19: Mann-Whitney U Test

$U = 280.5$, $Z = -2.89$, $p = .004$, $r = -.37$

Chapter 20: Kruskal Wallis H Test

$\chi^2 = 17.18$, $df = 2$, $p < .001$
Post hoc Mann Whitney tests show differences between groups 1 and 3 and groups 2 and 3.

Chapter 21: Spearman Correlation

Significant relationship:
Flutter tongue—Double tongue: $r_s = .54$, $p < .001$

Chapter 22: Wilcoxon Test

Wilcoxon $Z = -3.76$, $p < .001$, $r = -.58$

Chapter 23: Friedman's Test

$\chi^2 = 33.27$, $df = 4$, $p < .001$
Post hoc Wilcoxon tests demonstrate the significant differences to be between observations 3 and 4 and between observations 4 and 5.

REFERENCES

Abril, C. R., & Gault, B. M. (2006). The state of music in the elementary school: The principal's perspective. *Journal of Research in Music Education, 54*(1), 6–20.

Abril, C. R., & Gault, B. M. (2008). The state of music in secondary schools: The principal's perspective. *Journal of Research in Music Education, 56*(1), 68–81.

Bae, S., Zhang, H., Rivers, P. A., & Singh, K. P. (2007). Managing and analysing a large health-care system database for predicting in-hospital mortality among acute myocardial infarction patients. *Health Services Management Research, 20*(1), 1–8.

Ballantyne, J., & Packer, J. (2004). Effectiveness of preservice music teacher education programs: Perceptions of early-career music teachers. *Music Education Research, 6*(3), 299–312.

Bauer, W. I., Reese, S., & McAllister, P. A. (2003). Transforming music teaching via technology: The role of professional development. *Journal of Research in Music Education, 51*(4), 289–301.

Berg, M. H., & Austin, J. R. (2006). Exploring music practice among sixth-grade band and orchestra students. *Psychology of Music, 34*(4), 535–558.

Bright, J. (2006). Factors influencing outstanding band students' choice of music education as a career. *Contributions to Music Education, 33*(2), 73–88.

Clementson, C. J. (2014). *A mixed methods investigation of flow experience in the middle level instrumental music classroom* (Doctoral dissertation). University of Minnesota.

Costello, A. B., & Osborne, J. W. (2005). Best practices in exploratory factor analysis: Four recommendations for getting the most from your analysis. *Practical Assessment, Research & Evaluation, 10*(7), 1–9.

Davidson, J. W., Moore, D. G., Sloboda, J. A., & Howe, M. J. (1998). Characteristics of music teachers and the progress of young instrumentalists. *Journal of Research in Music Education, 46*(1), 141–160.

de l'Etoile, S. K. (2001). An in-service training program in music for child-care personnel working with infants and toddlers. *Journal of Research in Music Education, 49*(1), 6.

Ekholm, E. (2000). The effect of singing mode and seating arrangement on choral blend and overall choral sound. *Journal of Research in Music Education, 48*(2), 123–135.

Ester, D. (2009). The impact of a school loaner-instrument program on the attitudes and achievement of low-income music students. *Contributions to Music Education*, *36*(1), 53–71.

Gudmundsdottir, H. R. (2010). Pitch error analysis of young piano students' music reading performances. *International Journal of Music Education*, *28*(1), 61–70.

Guilbault, D. M. (2004). The effect of harmonic accompaniment on the tonal achievement and tonal improvisations of children in kindergarten and first grade. *Journal of Research in Music Education*, *52*(1), 64–76.

Hillier, A. J., Greher, G., Poto, N., & Dougherty, M. (2011). Positive outcomes following participation in a music intervention for adolescents and young adults on the autism spectrum. *Psychology of Music*, 0305735610386837.

Hogenes, M., van Oers, B., Diekstra, R. F., & Sklad, M. (2015). The effects of music composition as a classroom activity on engagement in music education and academic and music achievement: A quasi-experimental study. *International Journal of Music Education*, 0255761415584296.

Hopkins, M. T. (2002). The effects of computer-based expository and discovery methods of instruction on aural recognition of music concepts. *Journal of Research in Music Education*, *50*(2), 131.

Huberty, C. J., & Olejnik, S. (2006). *Applied MANOVA and discriminant analysis* (Vol. 498). John Wiley & Sons.

Kinney, D. W. (2008). Selected demographic variables, school music participation, and achievement test scores of urban middle school students. *Journal of Research in Music Education*, *56*(2), 145–161.

Könings, K. D., Brand-Gruwel, S., & van Merriënboer, J. J. G. (2007). Teachers' perspective on innovations: Implications for educational design. *Teaching and Teacher Education*, *23*, 985–997.

Kostka, M. J. (2000). The effects of error-detection practice on keyboard sight-reading achievement of undergraduate music majors. *Journal of Research in Music Education*, *48*(2), 114–122.

Levy, J., Kent, K. N., & Lounsbury, J. W. (2009). Big Five personality traits and marching music injuries. *Medical Problems of Performing Artists*, *24*(3), 135–140.

Lien, J. L., & Humphreys, J. T. (2001). Relationships among selected variables in the South Dakota all-state band auditions. *Journal of Research in Music Education*, *49*(2), 146–155.

Madsen, C., & Goins, W. E. (2002). Internal versus external locus of control: An analysis of music populations. *Journal of Music Therapy*, *39*(4), 265–273.

Martínez-Castilla, P., & Sotilla, M. (2008). Singing abilities in Williams syndrome. *Music Perception*, *25*(5), 449–469.

May, L. (2003). Factors and abilities influencing achievement in instrumental jazz improvisation. *Journal of Research in Music Education*, *51*(3), 245–258.

McCambridge, K., & Rae, G. (2004). Correlates of performance anxiety in practical music exams. *Psychology of Music*, *32*(4), 432–439.

McClung, A. C. (2008). Sight-singing scores of high school choristers with extensive training in movable solfège syllables and Curwen hand signs. *Journal of Research in Music Education*, *56*(3), 255–266.

Miksza, P., Roeder, M., & Biggs, D. (2010). Surveying Colorado band directors' opinions of skills and characteristics important to successful music teaching. *Journal of Research in Music Education, 57*(4), 364–381.

Morrison, S. J., Montemayor, M., & Wiltshire, E. S. (2004). The effect of a recorded model on band students' performance self-evaluations, achievement, and attitude. *Journal of Research in Music Education, 52*(2), 116–129.

Napoles, J. (2007). The effect of duration of teacher talk on the attitude, attentiveness, and performance achievement of high school choral students. *Research Perspectives in Music Education, 11*(1), 22–29.

Paul, S. J., Teachout, D. J., Sullivan, J. M., Kelly, S. N., Bauer, W. I., & Raiber, M. A. (2001). Authentic-context learning activities in instrumental music teacher education. *Journal of Research in Music Education, 49*(2), 136–145.

Pellegrino, K., & Russell, J. A. (2015, Spring). String teachers' practices and attitudes regarding their primary string instrument in settings inside and outside the classroom. *Bulletin of the Council for Research in Music Education* (204), 9–26.

Peynircioglu, Z. F., & Ali, S. O. (2006). Songs and emotions: Are lyrics and melodies equal partners? *Psychology of Music, 34*(4), 511–534.

Russell, J. A. (2006). The influence of warm-ups and other factors on the perceived physical discomfort of middle school string students. *Contributions to Music Education, 33*(2), 89–109.

Russell, J. A. (2008). A discriminant analysis of the factors associated with the career plans of string music educators. *Journal of Research in Music Education, 56*(3), 204–219.

Russell, J. A., & Austin, J. R. (2010). Assessment practices of secondary music teachers. *Journal of Research in Music Education, 58*(1), 37–54.

Russell, J. A., & Hamann, D. L. (2011). The perceived impact of string programs on K–12 music programs. *String Research Journal, 2*, 49–66.

Sheldon, D. A. (2004). Effects of multiple listenings on error-detection acuity in multivoice, multitimbral musical examples. *Journal of Research in Music Education, 52*(2), 102–115.

Siebenaler, D. (2008). Children's attitudes toward singing and song recordings related to gender, ethnicity, and age. *Update: Applications of Research in Music Education, 27*(1), 49–56.

Smith, B., & Barnes, G. V. (2007). Development and validation of an orchestra performance rating scale. *Journal of Research in Music Education, 55*(3), 268–280.

Stambaugh, L. A., & Demorest, S. M. (2010). Effects of practice schedule on wind instrument performance: A preliminary application of a motor learning principle. *Update: Applications of Research in Music Education, 28*(2), 20–28.

Stamou, L. (2010). Standardization of the Gordon primary measures of music audiation in Greece. *Journal of Research in Music Education, 58*(1), 75–89.

Teachout, D. (2004). Incentives and barriers for potential music teacher education doctoral students. *Journal of Research in Music Education, 52*(3), 234–247.

Teague, A., Hahna, N. D., & McKinney, C. H. (2006). Group music therapy with women who have experienced intimate partner violence. *Music Therapy Perspectives, 24*(2), 80–86.

Wehr-Flowers, E. (2006). Differences between male and female students' confidence, anxiety, and attitude toward learning jazz improvisation. *Journal of Research in Music Education, 54*(4), 337–349.

Wolfe, D. E., & Noguchi, L. K. (2009). The use of music with young children to improve sustained attention during a vigilance task in the presence of auditory distractions. *Journal of Music Therapy, 46*(1), 69–82.

INDEX

a priori, 24, 78, 95, 210
additivity, 240
adjusted means, 163, 175, 182, 188
adjusted r^2, 199
alpha, 18–19, 24
alternative hypothesis, 5
   in ANCOVA, 169
   in ANOVA, 100
   in chi Square, 259
   in dependent samples t-test, 71
   in factorial ANOVA, 119
   in Friedman's test, 310
   in independent samples t-test, 84
   in Kruskal Wallis H test, 284
   in MANCOVA, 182
   in Mann-Whitney U test, 272
   in MANOVA, 136
   in one-sample t-test, 62
   in Pearson correlation, 51
   in regression analysis, 198
   in repeated measures ANOVA, 152
   in Spearman correlation, 294
   in Wilcoxon test, 301
ANCOVA, 163
ANOVA, 93
assumptions
   in ANCOVA, 165–166
   in ANOVA, 95–96
   in chi Square, 258
   in data reduction, 211
   in dependent samples t-test, 69
   in descriptive statistics, 34
   in discriminant analysis, 228
   in factorial ANOVA, 117
   in Friedman's test, 308
   in independent samples t-test, 80–81
   in Kruskal Wallis H test, 280
   in MANCOVA, 179
   in Mann-Whitney U test, 270
   in MANOVA, 131–133
   in one-sample t-test, 60
   in Pearson correlation, 50
   in reliability analysis (Cronbach's Alpha) 240
   in reliability analysis (split-half) 247–248
   in repeated measures ANOVA, 150
   in Spearman correlation, 292
   in Wilcoxon test, 300
asymptotic tails, 10

Bartlett's test of sphericity, 212, 215, 221, 319
beta weight, 197, 198, 199
between group sum of squares, 104
bimodal, 9
bivariate regression, 194
Bonferroni adjustment, 18, 19, 99, 188, 190, 287, 289, 313, 315

Bonferroni post hoc, 165
Box M test
  in discriminant analysis, 226, 228, 234–235
  in MANCOVA, 179, 185–186
  in MANOVA, 133–134, 138
  in repeated measures ANOVA, 150–151

categorical, 21–22
causation, 47, 291
central tendency, 8, 14, 31, 40, 42, 296
chi-square test, 227, 248, 249, 250, 253, 257, 258, 259, 263, 264, 265, 266
classification results, 232
Cohen's d, 76, 88–90
confidence interval, 64
confirmatory factor analysis, 210
confounding variables, 165
conservative, 18, 19, 72, 107, 134, 151, 157, 199, 287
consistency, 69, 81, 96, 117, 218, 239, 240, 241, 242, 243, 245, 253
continuous, 21–22
control, 18, 19, 30, 83, 98, 114, 121, 124, 163, 164, 165, 167, 168, 176, 177, 180, 181, 190, 194, 283, 322
correlation matrix, 55, 215, 218, 221
covariate, 84, 135, 163, 164, 166, 167, 168, 169, 170, 171, 172, 174, 175, 176, 177, 180, 181, 182, 183, 184, 190
Cramer's V, 260, 263, 264, 266
Cronbach's alpha, 239, 240, 241, 244, 245, 320
cross– tabulation, 260
crossloadings, 217

degrees of freedom, 14–16, 25–26
dependent variable, 20–23
dependent-samples t-test, 28, 67, 69, 70, 71, 74, 75, 76, 78, 81, 147, 148, 153, 299, 300, 301, 307
descriptive statistics, 31, 34, 43, 137, 142, 171, 184, 317
dichotomous, 28
direct correlation, 47, 48, 291, 292

directional hypothesis, 5–6
discriminant analysis, 28, 30, 225, 227, 228, 229, 230, 231, 233, 234, 235, 319
dispersion, 11, 16, 150
dummy variables, 36
Dunn's multiple-comparison procedure, 280
Dunnet, 122

effect size, 24–25
eigenvalue, 140, 209, 215, 219, 226, 231, 232, 234, 235, 320
empirical rule, 17
error, 18–19, 22–23
error variance, 122
eta squared, 109, 126, 127, 138, 140, 141, 144, 150, 158, 172, 186, 187
explanatory variable, 194
exploratory factor analysis, 210, 219

F statistic
  in ANOVA, 103, 106
  in factorial ANOVA, 115
  in Levene for independent samples t-test, 82
  in MANOVA, 131, 140–141
F Table, 106
factor analysis, 207–208
factor loadings, 209
factorability, 211
factorial ANOVA, 113–114
factors, 18, 99, 125, 135, 142, 182, 204, 207, 208, 209, 210, 211, 212, 214, 215, 218, 219, 221, 225, 226, 227, 229, 240, 309, 323
fixed factor, 121, 124, 137, 141, 170, 184
formula, xvii-xix
  Bonferroni adjustment, 18
  Chohen's d, 76
  Cramer's V, 264
  dependent samples t-test, 75
  line of best fit, 197
  Mann-Whitney U effect size (r) 274
  mean, 8

null hypothesis, 4
phi, 264
skewness, 12
standard deviaiton of a population, 14
standard deviation of a sample, 14
standardized regression, 197
unstandardized regression, 197
Wilcoxon effect size (r) 304
frequency tables, 21
Friedman's test, 307–308

Games-Howell post hoc test, 107, 122, 142, 144, 145, 318
Gaussian distribution, 10
generalizability, 31
goodness of fit, 249
Greenhouse-Geisser adjustment, 151
Guttman split-half reliability coefficient, 239, 250

heuristic, 209, 211, 218, 236
hierarchical multiple regression, 194
histogram, 70, 96, 117
homogeneity of the covariance matrices, 133, 179
homogeneity of variance, 69, 81, 82, 96, 102, 103, 106, 117, 122, 166, 171, 172
homoscedasticity, 50, 51, 53, 196, 197, 228
Hotelling's T^2, 248, 249, 252, 253
Hotelling's trace, 139, 140, 186
Huynh-Feldt adjustment, 151
hypothesis testing, 3
hypothesis zero, 4
hypothesized mean, 59

independence of observations, 147
independent variable, 28
independent-samples t-test, 79
indirect correlation, 48, 292
individual difference variables, 7, 168, 284
inferential, 31, 34, 37–38, 43, 55, 69–70, 78–79, 81–82, 86, 96–97, 117, 166, 191, 271, 282, 301

interaction-effect hypothesis, 119
interactions, 113, 114, 126, 130, 132, 144, 166, 182
intercept, 197
internal consistency, 241
inter-rater reliability, 239
interval, 21, 22

Kaiser-Myer-Olkin (KMO), 211
kappa, 239
Kolmogorov-Smirnov test, 36
Kruskal Wallis H test, 279
kurtosis, 10–11, 34, 36–37, 68, 70, 117, 133, 150

lambda
  in discriminant analysis, 226–227, 232, 234–235
  in MANCOVA, 186–187
  in MANOVA, 131, 138, 139, 140, 144
leptokurtic, 11, 27
Levene test
  in ANCOVA, 166, 171
  in ANOVA, 96, 103, 106, 107
  in Dependent Samples t-test, 75
  in Factorial ANOVA, 118, 122, 127
  in Independent Samples t-test, 82, 88
  in MANCOVA, 187, 190
  in MANOVA, 133–134, 138, 140, 145
  in Repeated Measures ANOVA, 150–151, 157
liberal, 18, 72, 151, 157
line of best fit, 48, 50, 55, 197, 240, 248
linearity, 133, 179, 196, 211
logistic regression, 194

main effects
  in ANCOVA, 171, 182, 184
  in factorial ANOVA, 114, 115, 116, 118, 126, 127
  in MANOVA, 130, 132, 144
  in Repeated Measures ANOVA, 156, 157
main-effect hypotheses, 119
MANCOVA, 177
Mann-Whitney U, 269

MANOVA, 129
matched, 67, 299
Mauchly's test for sphericity, 150–151
Mauchly's W, 157
mean, 16, 31, 65, 126, 144, 150, 158, 289
mean gain scores, 271, 283, 285
mean rank, 274, 276, 286, 289, 310, 312, 315
mean square between groups, 105
mean square within group, 105
median, 8–10, 12, 31, 36–37, 40–43, 150
mesokurtic, 11
mode, 8–10, 12, 31, 37, 41–43, 150, 321
model development, 225
monotonic, 293, 294, 295
multicolinearity, 179, 196, 202, 204, 228
multimodal, 10
multiple regression, 194
multivariate analysis, 129, 132, 177

negatively skewed, 12, 37
nominal, 21, 38, 39, 60, 258, 260
nonparametric, 16, 22–23, 27–30, 34, 38, 96
normal distribution, 10, 13, 17, 27, 29, 37, 70, 82, 92, 96, 117, 166, 228, 300
normality, 12–13, 34, 36–37, 53, 82, 131, 133, 150, 179, 196, 228–229, 271, 300
null hypothesis, 3–7
　in ANCOVA, 168
　in ANOVA, 100
　in chi Square, 259
　in data reduction, 212
　in dependent Samples t-test, 70
　in discriminant Analysis, 228
　in factorial ANOVA, 118
　in Friedman's test, 310
　in independent samples t-test, 84
　in Kruskal Wallis H test, 284
　in MANCOVA, 182
　in Mann-Whitney U test, 272
　in MANOVA, 135
　in one-Sample t-test, 62
　in Pearson correlation, 51
　in regression analysis, 198
　in repeated measures ANOVA, 152
　in Spearman correlation, 294
　in Wilcoxon test, 301

oblimin rotation, 211
oblique rotation, 210, 214, 218, 219
oblique rotations, 210, 211
omnibus test, 108, 131, 140, 144, 178
one-sample t-test, 30, 63, 65
ordinal, 21–22
ordinate, 197
orthogonal, 207, 210, 211, 218, 219, 227
orthogonal rotations, 210
outcome variable, 20
outlier, 13
outliers, 23, 50, 53, 60, 86, 228, 240, 248

p value, 24, 88
paired-samples t-test, 67
parallel forms, 239, 247
parametric, 22–23, 29, 34, 48, 50, 60, 117
parametric Test, 30
parsimonious, 84, 93, 115, 129, 269, 279, 280
pattern matrix, 210, 215, 218, 221
Pearson correlation, 28, 30, 52, 291, 292, 294, 295
phi, 263, 264, 266
Pillai's trace, 139, 140, 186
platykurtic, 11
Poisson distribution, 13
policy significance, 25
positively skewed, 11, 12, 37
post hoc, ergo propter hoc, 57
post hoc tests
　in ANCOVA, 165
　in ANOVA, 94–95
　in factorial ANOVA, 126
　in Kruskal Wallis H test, 280, 287–288
　in MANCOVA, 179, 187
　in MANOVA, 131, 140–141
　in Repeated Measures ANOVA, 157
post-test, 67–72, 74, 77, 83–84, 164, 167, 174–176, 271, 299–303, 306
power, 19, 23, 86, 243
practical significance, 24–25
predicted variable, 20

predictor variable, 172, 193, 198, 204, 229
pre-experimental, 61
pre-test, 67–72, 74, 77, 83–84, 98, 164, 167, 175–176, 178, 181, 271, 283, 299–303
principal factor analysis, 208
principal axis factoring. *See* principal factor analysis
principal component analysis, 207
promax rotation, 211
probability, 24

quasi-experimental, 61

r^2, 199, 201
random sampling, 69, 81, 270, 300
range, 11, 14, 17, 36, 48, 131, 136, 185, 274, 286, 292, 302, 317
ratio, 21–22, 39, 60, 106, 140, 212, 215, 229, 292, 294
raw scores, 31, 197
regression, 193, 196, 197–200, 204–205, 319
regression coefficient, 193, 197
regression slopes, 166
repeated-measures ANOVA, 147–148, 150, 153–154
research questions, 7, 113, 147
response variable, 194
Roy's largest root, 139, 140, 186

scatterplots, 50, 53, 58, 179, 196, 243, 295
Scheffe, 107, 122, 125, 127, 142, 144–145, 165, 318
Shapiro-Wilk test, 36–37
simultaneous multiple regression, 194
skewness, 11–12, 36–37, 70, 96, 117, 133, 150, 228–229
Spearman correlation, 291, 292, 293, 294, 295, 296
Spearman's rank order correlation. *See* Spearman correlation
Spearman's rho. *See* Spearman correlation
Spearman-Brown coefficient, 239, 247, 250

sphericity, 150–151, 156, 159, 212, 214–215, 219, 221, 318–319
split-half reliability, 239
standard deviation, 12, 14–17, 26–27, 31, 34, 36, 39, 40
standard error, 31, 37, 63, 72, 74, 86, 174, 203, 229, 319
standardized canonical discriminant function coefficient, 232
statistical Package for the Social Sciences, 12, 35
statistically significant, 24, 25, 59, 90, 108, 187, 226, 234, 274
stepwise, 195, 196
stepwise multiple regression, 194
structure matrix, 210, 226, 232, 234, 235
sum of squares, 103, 104, 105, 106, 109, 116
symmetric measures table, 263

tolerance, 202, 203, 204, 319
total sum of squares, 103, 106, 109
t-table, 88
Tukey, 122, 165, 240, 242, 245, 247, 248, 249, 250, 252, 253, 320
Tukey's test of additivity, 242, 245, 253
Tukey's test of nonadditivity, 240, 247
two-tailed, 88
Type I error, 18, 19, 72, 76, 86, 99, 105, 113, 129, 151, 228, 287
Type II error, 18, 19, 22, 23, 72, 151

unimodal, 60
univariate tests, 129, 130, 132, 145
unstandardized regression formula, 197

variability, 14, 26, 228, 296
variance, 16–17, 23–24, 26–27, 31, 36, 39
varimax rotation, 211
VIF, 202, 203, 204, 319

Wilcoxon test, 299
Wilks's lambda, 131, 139, 186, 226, 227, 232, 234
within group sum of squares, 103

Z scores, 197